CONTESTED COUNTRYSIDE CULTURES

This book reflects the emerging interest in cultural aspects of rural studies. In rural society many social groupings are constructed as 'other' to the mainstream ideas of who belongs, or is welcome, in the contemporary countryside.

Contested Countryside Cultures examines this 'other' side of the countryside, a place inhabited by those marginalised, for example, by age, class, gender, race and disability. It seeks to question the processes whereby the cultural constructions of rurality not only appear to render the identities of these individuals and groups as 'other', but can also act to marginalise the people concerned.

The editors begin by discussing the notion of otherness. Here they warn that the concept should not be oversimplified – identity is not static and cannot easily be applied to distinct people groups – and that it raises some very important ethical questions for researchers. These themes are amplified in a series of conceptual essays in the first part of the book, in which different topographies of rurality and otherness are mapped out. The second part of the book offers a series of essays reflecting current research findings on the otherness and marginality associated with particular aspects of age, gender, sexuality, economic position and alternativeness.

The book makes public some conceptual and empirical advances in the understanding of how rural society is differentiated in line with cultural constructions of rurality. It also reflects, however, the need to stay in touch with issues raised by other approaches, notably the contribution of political economy to the understanding of socio-economic marginalisation.

Paul Cloke is Professor in the Department of Geography at the University of Bristol. **Jo Little** is a senior lecturer in the Department of Geography at the University of Exeter.

CONTESTED
COUNTRYSIDE
CULTURES

Otherness, marginalisation and rurality

Edited by Paul Cloke and Jo Little

London and New York

First published 1997
by Routledge
11 New Fetter Lane, London EC4P 4EE

Simultaneously published in the USA and Canada
by Routledge
29 West 35th Street, New York, NY 10001

Phototypeset in Garamond by Intype London Ltd
Printed and bound in Great Britain by
Biddles Ltd, Guildford and King's Lynn

British Library Cataloguing in Publication Data
A catalogue record for this book is available from the British Library

Library of Congress Cataloging in Publication Data
Contested countryside cultures: otherness, marginalisation, and rurality / edited
by Paul Cloke and Jo Little.
Includes bibliographical references and index.
1. Sociology, Rural. 2. Rural conditions. 3. Marginality, Social. 4. Group
identity. I. Cloke, Paul J. II. Little, Jo
HT421.C64 1996
307.72–dc20 96–43171

ISBN 0–415–14074–9 (hbk)
ISBN 0–415–14075–7 (pbk)

CONTENTS

v

CONTENTS

vi

ILLUSTRATIONS

FIGURES

TABLES

CONTRIBUTORS

Julian Agyeman	School of Urban Development and Policy, South Bank University
David Bell	School of Arts, Staffordshire University
Paul Cloke	Department of Geography, University of Bristol
Clare Fisher	Department of Geography, University of Wales, Lampeter
Keith Halfacree	Department of Geography, University of Wales, Swansea
Sarah Harper	Wellcome Unit for the History of Medicine, Oxford
Annie Hughes	School of Geography, Kingston University
Owain Jones	Department of Geography, University of Bristol
Jo Little	Department of Geography, University of Exeter
Jonathan Murdoch	Department of City and Regional Planning, University of Wales, Cardiff
Chris Philo	Department of Geography and Topographic Science, University of Glasgow
Andy C. Pratt	Department of Geography, London School of Economics and Political Science
David Sibley	School of Geography, University of Hull
Rachel Spooner	Department of Geography, University of Bristol
Gill Valentine	Department of Geography, University of Sheffield

1

INTRODUCTION: OTHER COUNTRYSIDES?

Paul Cloke and Jo Little

OTHER COUNTRYSIDES?

David Sibley in his excellent book *Geographies of Exclusion* (1995) quotes from a speech by the Conservative MP Sir Cranley Onslow during the second reading of the Criminal Justice and Public Order Bill (1994):

> Part V [the section of the bill dealing with public order differences] and its provisions to strengthen the position of those people who want law and order to prevail in the countryside are an important departure from precedent. The creation of a new offence of aggravated trespass is a significant step forward that will be widely welcomed in all parts of the country where people have become all too used to disorder, intimidation and violence prevailing and interrupting the lawful pursuits of those who live in the country, value it and want to continue with their countryside sports.
>
> (Sibley 1995: 107)

Here, Sibley notes, is one blatant vision of rural life which clearly excludes a host of 'others' from a supposed countryside hegemony based on idyllic villages, red-coated huntsmen and a certain conservative Englishness. We wanted to call this book 'Other Countrysides' because we believe that representations of rurality and rural life are replete with such devices of exclusion and marginalisation by which mainstream 'self' serves to 'other' the positioning of all kinds of people in the socio-spatial relations of different countrysides. Instead, our title *Contested Countryside Cultures* must serve this purpose, and our hope is that 'contesting' will signify *both* a wish to contest any generalised rural other, *and* the inevitability that different representations of culturally constructed countrysides will be mutually contesting in one way or another.

This book reflects something of a resurgence in rural studies over recent years, not only as it embraces the 'cultural turn' which is evident in the broader social sciences, but also as it achieves a wider significance, as the cultures of nature and rurality are (re)discovered to be transcending the

1

supposed boundaries of rural geographical space. In particular, three foci have emerged which have bridged the gulf between rural studies and what have previously been supposed to be more mainstream cultural ideas. First, the focus on *landscape* has involved a recasting of cultural geography's traditional interest in landscape and environmental relations. The recognition that environmental, territorial and other geographical myths are of continuing significance in constructing alternative nationalist ideologies (see, for example, Daniels 1993) has served to make interesting connections between countrysides and ideological representations. Second, the focus on the *spatiality of nature* has implicated countryside spaces in increasing measure. Given the rising interest in the relations and agency of nature and environment, countrysides are often represented as appropriate and obvious spatialisations of nature in which to contextualise wider themes. Third, there has been a focus on *'hidden' others* which is the principal concern of this book. Here, rural areas have proved to be a magnetic attraction for those seeking to practise cultural geography's theorisations of difference and otherness. Countrysides are rich in myth, and they offer a scale of territory which somehow provides researchers with a heightened prospect of access to hidden others in appropriate numbers. Here too, then, rural studies have bridged over into wider concerns, with considerable intellectual excitement being generated in the process.

Identification of these foci should not be taken to suggest that the only geographies of rural areas are now these cultural ones. Indeed, we would suggest that these more recent cultural geographies are being overlaid, palimpsestually, onto existing accounts of behavioural and political studies of economic change, and constructionist studies of social change. We do suggest, however, that some of the current excitement and 'fizz' of rural studies does draw heavily on these cultural themes. There has been a move from old to new centralities, involving a change to the 'stuff' of rural studies, both theoretical and methodological (Cloke 1997). We have no wish at this point to reproduce existing accounts of the changing nature of rural studies (see, for example, Cloke and Goodwin 1993; Marsden *et al.* 1993; Phillips 1997), but we do wish to comment briefly on the broad trajectory of these changes.

In terms of theoretical 'stuff', rural studies has travelled a tortuous and non-linear journey from a fascination with theorising regularity, via a fascination with a critical theorising of the sameness inherent in the structuring of opportunities and the agency of human decision-making, to the more recent emphasis on theorising differences and significations in geographies of otherness, discourse and cultural symbolism. There are currently some very interesting concurrent spirals in this work, with different strands interweaving not only between different people's work but also within some people's work. Thus, attempts to explain rural change are spiralled together with a giving voice to other geographies of rural life and change; relativist attitudes to power relations are spiralled with ideo-logical attitudes to power; and geographies of (so-called) 'real' life in rural settings are spiralled with imagined geographies of the rural.

The shift from old to new centralities is also manifest in methodological practices, as rural researchers and scholars have moved from a fetishism with numeric data towards the interpretation of a kaleidoscope of different texts. There is now a more marked fascination with the imaginary texts of novels, paintings, photographs, films, television and radio. In many ways, images and stories about countrysides have proved very alluring, with the very idyll-ised myths about nature and rural life (which are often the focus of deconstruction) drawing researchers to themselves in an obvious, yet ironic, process by which research subjects are chosen. The willingness to embrace the stories told in these imaginary texts is inextricably linked with the methodological fascination with ethnographies. Constructing and interpreting qualitative texts has often involved a facilitation of story-telling by research subjects. Increasingly easy transitions can thus be made between the 'imagined geographies' of imaginary text, and the 'imagined geographies' of rural people. An integral component of these new rural ethnographies has been the desire to give voice to 'other' rural geographies, although there has inevitably been a concurrent concern for the intertex-tualities of the situated author-knowledges of the self.

This book, then, seeks to present some of the fruits of these new fascinations. However, it does not do so uncritically. 'Others' as positioned subject groups or individuals and 'othering' as processes by which such positioning occurs are certainly not unproblematic ideas. Therefore, before proceeding to some of the stories that have been assembled regarding rural others, we want to look a little further at the concepts which for some are viewed as reinvention of very conventional concerns for inequality, discrimination and social division, but which for others open new windows onto the social and spatial processes of boundary formation in rural areas whereby some groups and individuals are separated out from society as being different, and often deviant.

OTHER CONCERNS?

The focus on 'otherness' in rural studies has been signposted by a recent debate in *Journal of Rural Studies*. Philo's review of neglected rural geo-graphies emphasised that most accounts of rural life have viewed the mainstream interconnections between culture and rurality through the lens of typically white, male, middle-class narratives:

> there remains a danger of portraying British rural people ... as all being 'Mr Averages', as being men in employment, earning enough to live, white and probably English, straight and somehow without

sexuality, able in body and sound in mind, and devoid of any other quirks of (say) religious belief or political affiliation.

(Philo 1992: 200)

Here, Philo not only highlights some 'forgotten' items for the rural research agenda but also points to the discursive power by which mythological commonalities of rural culture will often represent an exclusionary device, serving to marginalise individuals and groups of people from a sense of belonging to, and in, the rural, on the grounds of their gender, age, class, sexuality, disability and so on. As rurality is increasingly understood as a phenomenon which is socially and culturally constructed, so the exclusionary qualities within these constructions need to be highlighted. To offer an example which is worked through in Chapter 14, if rural myths of problem-free living preclude recognition of poverty (Cloke 1995; Cloke *et al.* 1995) or homelessness, then we need to know about it. Or, as discussed in Chapter 11, if rurality is bound up by nationalistic ideas of (white) Englishness then resultant cultural attitudes about who does and who does not belong in the countryside serve as discriminating mechanisms of exclusion. Power is thereby bound up discursively in the very socio-cultural constructs which have characterised rurality.

Philo's debate with Murdoch and Pratt hinges on the degree to which understandings, evaluations and uses of understandings about the discursive power within socio-cultural constructions of the rural can be channelled collectively into an agenda for change. He suggests that rural geography should be made more open to the 'circumstances and to the voices of "other" people in "other" places: a new geography determined to overcome the neglect of "others", which has characterised much geographical endeavour to date' (Philo 1992: 194). Murdoch and Pratt's response is that merely giving voice to 'others' does not sufficiently address issues of power in the rural arena:

what follows from this concern to 'give voice' are a set of issues which Philo does not really consider ... Simply 'giving voice' to 'others' by no means guarantees that we will uncover the relations which lead to marginalisation or neglect.

(Murdoch and Pratt 1993: 422)

Philo's thesis on power and 'otherness' is founded upon a postmodern rejection both of transcendental notions of the problematic, and of a priori theorisations of injustice. He admits to being

unhappy about the assertive modernist impulse present in Bauman (and thus in Murdoch and Pratt) which proceeds with such certainty, which still puts faith in the a priori theoretical specification of how the world and its injustices operate, and which heroically assumes the duty of assessing from without the realities of 'other lives' against

4

transcendental yardsticks of 'right'/'wrong', and 'good'/'bad' that may have little relevance for the peoples and places concerned.

(Philo 1993: 433)

Murdoch and Pratt's reply underlines important cuts of difference in the motives which inform and spark the imagination of rural research. In particular, there are important differences between those wishing to allow particular people in particular places to speak for themselves about the power relations in which they are located, and those wishing to incorporate a vision for change into the research process:

> Should we not attempt to reveal the ways of the 'powerful', exploring the means by which they make and sustain their domination (perhaps in the hope that such knowledge could become a reservoir to be drawn upon by oppositional actors)? Do we not also seek to influence the decisions of the 'powerful', such as policy-makers, in the hope that they might be persuaded to produce more effective and just interventions in the world?

(Murdoch and Pratt 1994: 85)

Far from being a comfortable concept, then, otherness is a focus of enquiry which can be driven from very different philosophical standpoints and which carries with it very different expectations in relation to the politics of 'giving voice' to othered individuals and groups.

A sensitive understanding of 'others' suggests an acknowledgement of several complex issues. First, to talk of 'other' is to gaze into a mirror of the self. As Shurmer-Smith and Hannam explain, not only do we have to escape the tunnelled vision of our own gaze on the world, but we also have to realise that in so doing we risk the homogenisation of the non-self:

> all of us are constantly falling into the trap of assuming that our own view is obvious, no matter how hard we try to think beyond our own egocentricity, ethnocentricity or group solidarities. But as soon as we start to think about people who are not ourselves we lapse into the language of 'Othering' and, as one urges oneself to consider 'Others' or to see the 'Other' side of the question, those who are not like 'me' can start to slide into homogeneous mass of difference from 'me', *essentially* the same as each other.

(Shurmer-Smith and Hannam 1994: 89)

By this token, the self dominates, and the other comes a very poor second, a tolerated periphery marginalised by individualist politics and often painted out in the canvas we use to impose knowledgeable order onto a hugely variegated world. Recognition of the need to give voice to neglected others therefore requires not only a deconstruction of these

knowledge–order frameworks, but also a full recognition of the intertex-
tualities of the self. Our natural tendency is to impose familiarity of
language, concept and representation onto other subjects and thus to treat
them as what Doel (1994) labels as 'the other of the same'. The alternative
(Doel's 'other of other') is an openness to new categories, unfamiliar
interconnections and unknown language concept and representation. As
Philo suggests in Chapter 2, the terrain of rural studies has thus far been
characterised by the other of the same rather than the other of the other.

If other is inextricably linked to the self, then it is also to be understood
in terms of an unrecognised positioning of subjects. Lacan's (1977) conceptualisation of othering in terms of child development traces a
psychoanalytical division between those aspects of the self which are central
and those which are peripheral. Just as a child looking in a mirror sees the
self that is looking and the self that is looked at, so the more general
constitution of self and subject involves a phalanx of meanings which
reinforce an automatic positioning of subjects as core or peripheral. These
constitutive meanings tend to be implicit such that the positioning of the
subject tends to be essentialised rather than the reflexive result of assessing
the power structures at work.

A third issue in an understanding of others relates to the domain in
which 'othering' is experienced. Sibley (1995) shows how otherness is
sometimes thought to reside in some separate domain of text or psyche.
For example, Deleuze (1990) offers a construction of the other in which
the self is experienced as an imaginary being, where places, ideas, sensations
which are not directly experienced can achieve an imagined existence.
Following Mead (1934) however, Sibley prefers an interpretation of the
relationship between the self and the other which is fully located in
the social and material world. Thus the boundaries between the self and the
other may be formed through a series of cultural representations of people,
positionings and objects, but are embedded in social and often spatial
contexts.

Some of these complexities are evident in the practising of research into
ideas of otherness in specific circumstances. Most contemporary studies of
otherness draw in some degree on the accounts of Orientalism by Said
(1978, 1986, 1993). Here, the relationship between the Orient and the
Occident is traced to a created nexus of theory and practice in which
the discursive formation of the Orient as 'other' rests on a complex
hegemony of domination which has been invested in both materially and
culturally by the colonial powers of the West. Said's work has been seminal
in its critical focus on the discursive constitution of colonialism which
deploys a powerful criticism of the self, and an equally powerful appreciation of other discourses of the Orient. As Bernstein (1992) has stated in
relation to Derrida's practices of deconstruction, the power of this work
lies in 'an uncanny ability to show us that at the heart of what we take to

be familiar, native, at home – where we think we can find our center – lurk (is concealed and repressed) what is unfamiliar, strange and uncanny' (p. 174). There is certainly a sense in which the familiar, natural belonging so often associated with the home and hearth of rurality has lurking within it relations and positionings which are unfamiliar, strange and literally uncanny. It might also be suggested that rurality is subject to forms of internal colonialisation not unlike those described by Said, in that the discursive formation of the rural rests on a complex hegemony of domination which both materially and culturally constitutes an acceptance and belonging for some, and a marginalisation and exclusion for others. What is clear, is that the complexities of rural selves and rural others will continue to be a key arena in the appreciation of the interrelations of socially and culturally constructed ruralities.

As with all such prominent contributions, Said's studies have themselves attracted considerable criticism from those wishing to practise research into ideas of otherness. For example, Lewis (1993) and Rocher (1993) have suggested that Said's concept of Orientalism is both too inclusive, and not inclusive enough. As Jewitt (1995) explains, the work is 'too inclusive' because it lumps together very different Anglo-French commentaries on the Orient, and thereby conflates very different voices, grammars, attitudes and depictions of their so-called others. However, it is 'not inclusive enough' because it tends to depict Orientalism only as the imagination of territorial imperialism, leading perhaps to the charge that the critique of material colonialism can become so relativised that nothing practical emerges from it. Such critiques offer useful warnings for those wishing to identify, give voice to, and even emancipate those who are positioned as 'other' in rural life and territory. Any seemingly hegemonic power relations will be complex and multi-faceted, defying attempts to attribute discourses of otherness to particular sources, and networks of cultural circulation. It is easy to be too inclusive in this respect. However, it may also be too simplistic merely to ascribe rural selves and others to the territory of the imagination, without interconnecting the imagined rural to include practices, processes and relations of discrimination, marginalisation and exclusion. We should also acknowledge that these positioned interconnections are not static. Morley and Robbins (1995) stress that the interrelations of space and identity occur in fluid and dynamic systems of relations and representations, such that person/identities can be variously positioned in different times and spaces. Subjects, selves and others may thereby be thought of as hybrid and may not easily be located in obvious domains of difference, such as race, gender or age. Indeed, Pile and Thrift (1995) urge us to look 'in between' these domains to discover the ongoing processes of negotiation and renegotiation by which selves and others are represented. Equally, it will be important to recognise that identity will be constructed differently through time, and that the reflexive presentation of the self will thereby not be contained within easily identifiable categories. Philo's (1993) call for attention to neglected rural geographies, then, should not simply be interpreted as an extension of rurality as positioned by gender, race, sexuality and so on – although such an extension is to be welcomed as a fruitful first step. The reversal of neglect should also account for other more hybrid geographies located in the interstices (or 'third space') between these categories.

OTHER IMPLICATIONS?

Thus far, in discussing some of the complexities inherent in the concept of 'other' rurals, we have restricted our attention to different emphases within broadly postmodern and poststructural approaches to otherness. There are, however, more far-reaching commentaries drawn from more oppositional philosophical perspectives. In general, such commentaries are critical of the politics of the postmodern which, as Schatzki (1993) comments, navigate between foundationalism and relativity and thereby induce an apprehension that politics can be reduced to 'recklessness, frivolity and paralysis' (p. 40).

One influential critic is Harvey (1989, 1990, 1993), who speaks from the position that relativist, essentialist and non-dialectical approaches to situatedness will inevitably generate immense political difficulties, of which at least four are specifically relevant to notions of otherness:

- Assuming that situatedness is a separate and unrelated difference creates representational difficulties for those who see the need to speak for 'others' and to speak out against the oppression of those whose identity is construed as other.
- A general and relativist respect for identity and otherness devalues attempts to suggest that 'although all others may be others, some are more other than others and that in any society certain principles of exclusion have to operate' (Harvey 1993: 63).
- If studies of otherness lead inexorably to a concern for the condition of identity politics, there may be an underemphasis on the relations between the social processes of construction of identities, and the condition of those identities.
- The very pursuit of identity politics may actually serve to perpetuate the processes by which certain identities are marginalised.

In sum, Harvey (1989) warns against the seductive nature of 'otherness' and fears that a focus on the other tends to 'draw a veil over real geography' (p. 87). Clearly 'real' here is a positioned evaluation, drawing on what Morris (1992) terms the mythic space of metatheory, and perhaps illustrating Gregory's (1994) assessment of an essential strength of postmodernism and postcolonialism, namely that 'in their different ways they can remind us of our own "otherness" ' (p. 409).

An appreciation of some of these broader critiques lead us to suggest five unresolved issues for the continuing study of otherness in the rural arena (Cloke 1997). The list is not exhaustive and is certainly interrelated, but can be used as a basis from which to ask questions of grounded studies of otherness and rurality.

Politics

The debate between Philo and Murdoch and Pratt mentioned earlier in this chapter raises important issues about why neglected rural others should be studied. Philo's (1992) account of the conventional gaze on rural issues highlights the assumption that British rural people are a kind of Mr Average. This gaze offers specific cuts on the othernesses involved, pointing to the relations and identities relating to gender, race, age, health, sexuality, alternativeness and so on. Merely to facilitate the emergence of other discourses representing these identities is thought by Murdoch and Pratt to ignore the pressing need to uncover the relations which lead to marginal-isation of these other identities. Is there a sense, then, that in moving away from metatheoretical notions of sameness, difference and right/wrong we have traded away a politics of conviction for a politics of identity where the perils of spatial relativism can lead to an 'anything goes' philosophy (Keith and Pile 1993)? Is there a danger that we trivialise vital issues such as gender, race and sexuality by 'othering' them within a relativist framework? As Swyngedouw (1995) has suggested, the loss of any consti-tutive contextualisation of politics and place can turn a commitment to emancipatory social practice and politics into a commitment to the political empowering of pleasure.

Policy discourse

Linked with these issues of politics, there are also some unresolved issues about the potential connectedness between research and public policy. In many ways, some contemporary rural research relating to issues of other-ness are discourse-incompatible with rural policy agencies and practices. An example close to the heart of one of us relates to the Rural Lifestyles research programme (Cloke et al. 1994, 1995, 1997) in which attempts were made to paint a picture of rural life in England in part by routing academic discourse through lay discourse (Halfacree 1993) so as to con-struct more qualitative and experiential narratives of the different needs, experiences and lifestyles of rural people. The texts constructed in this way not only encouraged the hearing of people's own voices, but also permitted an approach whereby the categories of interpretation could be structured under headings used by the research subjects themselves rather than conforming to 'obvious' policy-related headings. The result was a series of narratives which emphasised difference rather than sameness, which gave weight to 'unrepresentative' segments of interview text, and which did not offer easy linkages between narratives and policy solutions.

Two implications emerged from the presentation of this work to policy-making agencies (Cloke 1996b). First, there were obvious clashes with existing policy discourses, which were based on conventional headings and which valorised numerical texts. Second, the lifestyles research was vulnerable to a response from politicians and policy-making agencies based on a politics of individual responsibility. If there are no *clearly definable* problems in rural areas, we were told, then there is no need to bother about policy responses! Clearly a gulf of language and discourse can emerge between ethnographic studies of identities and positions of otherness and these policy terrains. The issue of bridging between these discourses could be a difficult one to resolve.

Morality

A further set of issues relating to why we research particular subjects using particular methods may be approached under the loose heading of morality. Bell's (1994) study of the Hampshire village 'Childerley' discusses two sorts of moral thinking displayed by residents. First, there is that moral thinking which Bell considers to be 'socially derived' in that social interests underlie what people say and believe as well as what they do. This way of thinking is regarded by Childerley residents as painful, and consequently is usually circumnavigated at every available opportunity. Secondly, there is that moral thinking which can be based on truths that can be represented as being above and free from the polluting interests of social life. In Childerley, this kind of thinking could be achieved by people conceiving of themselves and presenting themselves as 'country' or 'village' folk with the basic interests of the countryside or village at heart.

We consider that the idea of 'moral thinking' is equally valid when seeking to understand academic discourses. As with the people of Childerley so perhaps it is with rural researchers, as contemporary approaches and practices further permit a moral thinking which is above and free from social interest to supersede a moral thinking which is socially derived. If so, we can arrive at a rurality whose selves and others are morally constructed so as to avoid the social relations of place and the social and political promptings of the research imagination.

Privilege

One of the potential hazards of a focus on 'otherness' is that research will privilege particular forms of other to the neglect of other others. In the selection of rural subjects for research, it seems likely that both the broader fashions of the 'cultural turn' and the more place-specific richness of rural myths will present some subjects with more seductive appeal than others. Thus idyll-ised landscapes and places are favoured for study at the expense of less glamorous subjects and things which are woven within landscape tapestries. 'Ordinary' other places can become shadowed out by the privileging of special landscapes, with the result that the 'messiness' of rural space is sometimes lost (as indeed is the messiness of taken-for-granted metaphors such as text, landscape, mapping and so on).

Equally, it can be argued that we have privileged certain identities of otherness in seeking to rectify the neglected social geographies of rural life. An off-the-hip listing of otherness – gender, sexuality, race, age, disability, alternativeness and so on – has emerged which both risks the oversimplification of these very important and complex domains of difference, and serves to ignore other othernesses – those which are less easily categorised and perhaps hybrid, those which are less glamorous, those traits of otherness which may be a partial and transitory aspect of people's identity but which will often not be used to categorise the whole self, and so on – which contribute to positionings and identities that are important in the messiness of rural populations and their lifestyles.

Tourisms?

A final concern relates to the way in which we can sometimes privilege certain kinds of otherness without giving requisite attention to the importance of sustained, empathetic and contextualised research under clearly thought through ethical conditions. The tendency to 'flit in and flit out' of intellectually trendy subjects can have important implications in terms of the practice of research as mere tourism or voyeurism of the subjects concerned. This danger has been widely recognised by writers on the relations between feminism and postmodernism. For example, Bondi and Domosh (1991, 1992) suggest:

> there is a tendency within postmodernism to indulge in a kind of 'tourism', manifest in references to 'woman' and the 'feminine'; we have raised objections to what we see as an appropriation of women's voices and experiences by white male theorists.
>
> (Bondi and Domosh 1991: 211)

These objections are more widely applicable in that all of the 'otherness' typically identified by rural researchers can be similarly violated by research tourism. The problem here, however, is to find an appropriate balance between restricting research on particular subjects to 'non-tourists' (however defined) – which heightens exclusivity and indeed establishes a form of privilege in the research process – and encouraging ethically appropriate research on a range of identity-subjects for the positive reason of broadening and deepening understandings of connections between self and other which otherwise would perhaps not be made. This balance is problematic. Haraway (1990) has spoken of 'the very fine line between the appropriation of another's (never innocent) experience and the delicate construction of the just-barely-possible affinities, the just-barely-possible connections that might actually make a difference in local and global histories' (p. 241), and in our view it is possible to detect moods of playfulness and 'right-on-ness' in some recent work which would fall into Haraway's appropriation category, and which should be regarded as unfortunate or unacceptable depending on the political and ethical criteria being employed.

We make no claim here that the research and writing which constitute the contributions to this book are free from contention when set against these five criteria. Indeed, different authors vary in their reflection on them. Some confront them directly; others weave a more subject-based narrative in and around them. Our own view of these implications of the other are that they serve both as a warning against trivialising, de-sensitising and perpetuating othered identities and positionings, and as waymarkers to ethical, honest and reflexive attempts to deconstruct and reposition the mainstream selves of rural studies. Much depends on our individual answers to how we should encounter otherness in terms of a sensitive and contextualised identity politics rather than a more frivolous empowering of pleasure.

OTHER RURAL STUDIES

In identifying this recent and (we think) exciting focus on the 'hiddenness' of 'otherness' in rural studies, we find ourselves facing a number of important questions about appropriate methods of gaining access to the narratives of the other. Some of these issues have been debated over many years within academic discourses. We refer here to what Cohen (1994) describes as 'the characteristic anthropological problem' which poses 'unanswerable problems': 'How do you know *what* the other person is thinking? How do you know *that* the other person is thinking? How can you discriminate between the other person's consciousness and your construction of his or her consciousness?' (p. 3). Cohen's answers to these dilemmas allow him to establish a clear appreciation of what academic discourses (in this case in anthropology) represent and can achieve:

The answers to the first and second questions, 'I cannot know for certain', lead inexorably to the answer to the third: 'I cannot'. What we can do, what anthropologists customarily have done, is to use literary devices of one kind or another to convey in our authored texts the impression of such a discrimination. But it is one which we as authors have engineered.

(Cohen 1994: 3)

Cohen points us very clearly to some of the methodological inadequacies of the past and to the inevitability that interpretation of other selves will begin with our own selves (see also Cloke 1994). Our literary devices of authorship may appear to discriminate between the researcher self and the researched other, but in so doing we engineer difference rather than recognise the intertextualities of seeking access to other narratives.

Other questions have come to the forefront of academic discourse more recently, and they refer to the issue of the *authorisation* on which 'other' rural studies rest. Gregory (1994) poses these questions thus: 'By what right and on whose authority does one claim to speak for those "others"? On whose terms is a space created in which "they" are called upon to speak? How are "they" (and "we") interpellated?' (p. 205). Moreover, such questions not only provide pointers to how social scientists such as human geographers might approach their subjects, but they also (as Gregory insists) have 'geographies inscribed *within* them' (p. 205). Is it inevitable, then, that in seeking to access narratives of others and otherness we become mere tourists, weighed down by the dominant self-authority of author-power, and unable to escape the condemnation of our own sensibilities? Or are there ways in which we can sensitively acknowledge the intertextuality of our authorship and seek to understand the words of others without polluting them with tendencies to voyeurism? Gregory's answer is emphatic, and optimistic:

To assume that we are entitled to speak only of what we know by virtue of our own experience is not only to reinstate an empiricism: it is to institutionalize parochialism. Most of us have not been very good at listening to others and learning from them, but the present challenge is surely to find ways of comprehending those other worlds – including our relations with them and our responsibilities toward them – without being invasive, colonizing and violent. If we are to free ourselves from universalizing our own parochialisms, we need to learn how to reach beyond particularities, to speak of larger questions without diminishing the significance of the places and the people to which they are accountable. In doing so, in enlarging and examining our geographical imaginations, we might come to realize not only that our lives are 'radically entwined with the lives of distant strangers' [here Gregory quotes from Ignatieff 1984] but also that we bear a continuing and unavoidable responsibility for their needs in times of distress.

(Gregory 1994: 205)

In seeking to contest the hegemonic and mythical cultures of the countryside, which often serve to hide and marginalise a whole host of other identities, we are happy to echo Gregory's manifesto. In bringing together this collection of essays we do (perhaps too grandly) seek to enlarge and examine geographical imaginations of rurality, but in doing so we seek to avoid any abrogation of the responsibilities to the rural others which form the varied subject of our narratives.

Again, this positioning in our role as editors accounts perhaps for the shape of this book but does not restrict the imaginations of the authors of individual essays. The book's shape can be simply summarised. First there are a series of essays which variously reflect on aspects of otherness in the rural context. Chris Philo's essay (Chapter 2) focuses on the story of the Shaker Lane community, a fictional group of rural dwellers, placed somewhere in the rural USA, who display characteristics which differ from there found in orthodox accounts of rural life. He points us towards an appreciation of *different* other rural peoples and rural spaces, and in so doing warns us about the tendency in academic discourses to impose a recognisable 'sameness' even when researching 'other' subjects. In Chapter 3, Jon Murdoch and Andy Pratt map out the different topographies of rurality constructed in recent academic discourses, and trace the parallel contours of power which are often concealed but which usually sustain the more visible landscapes of research and scholarship. They argue for further attention to be paid to middle-level concepts (for example, region, network and fluidity or 'third space') by which rurality may be studied more fairly, and perhaps even more modestly. Keith Halfacree suggests in Chapter 4 that the shift towards post-productivist countrysides has opened up new spaces for the expression of new identities within the rural context. In his study of the attitudes and motivations of migrants, he notes that neo-tribal connections can form relational otherness in the countryside which can provide future openings for more radical polity visions for a more inclusive rural future. David Bell (Chapter 5) approaches the issue of rural otherness by discussing how American filmic representations of rurality demonstrate the flip side of popularly constructed rural idylls. Rural romanticism, he argues, has a hidden underbelly in the suffocating and repressive nature of rural small towns.

Second, the book offers a series of essays which focus on particular cuts on the issues relating to rural others, the connections and interstices between them, and the consequent questioning of orthodox representations of rural culture(s). It is not the intention here to provide a definitive listing of 'otherness' in the rural context. Neither do we suggest that the contributions here represent a prioritisation of the most important 'other' subjects in rural studies. Rather, the collection brings together what we considered to be some of the more interesting studies which had stories to tell when the book was mooted. The current excitement of rural studies means that many other interesting and important stories on these issues are emerging all the time.

In Chapter 6, Gill Valentine tells of attempts by lesbian feminists in the USA to construct a very different kind of rural idyll during the 1970s and 1980s. She describes the establishment of a series of 'lesbian lands' – separatist communities arranged in non-heteropatriarchal spaces – and shows how even these new sites of sameness inevitably generated new marginalised others, as some identities became valued more than others. Annie Hughes (Chapter 7) also describes the complexity of women's lives in the countryside, although her focus on the negotiation of identity by women in rural Wales offers interesting contrasts with the previous chapter. She concludes that women's domestic identities are complex and 'messy' as moral and social codes are criss-crossed by economic necessity and cultural affinity. Jo Little in Chapter 8 provides a detailed account of the connections between employment and the marginality of rural women. She argues for a far more sensitive reading of these issues, going beyond grand assumptions about labour and identity towards a more subtle appreciation of how gender identity, household strategy and labour market participation interact in the construction and experience of 'othered' gendered positionings.

In Chapters 9 and 10, the focus shifts to issues relating to age. Owain Jones explores some of the ways in which rural children are seen as 'other', and challenges the idyll-isation of country childhoods. In so doing he deconstructs the twin spirals of the imagined and remembered *adult* geographies of rural childhood, and the experience of rural life by children (which indeed will be at least in part constructed by those adult geographies). Sarah Harper examines the contradictions which confuse the construction of later life in rural areas. She argues that the older rural populations are not only constructed as a kind of self-evident other, but also that such othering interconnects in a complex way with the perceived marginality and othering of rural space.

Julian Agyeman and Rachel Spooner (Chapter 11) pose important questions about the role of rurality as a signifier of white national identity, and indicate the nature of racist attitudes towards people of colour who live in, or might want to visit, the countryside. They trace the interconnections between rurality and ethnicity, charting the potential for contesting the 'white space' culture of the countryside, particularly in terms of the work of the Black Environmental Network to encourage people of colour to became involved with environmental issues. In Chapter 12, David Sibley similarly recognises the 'sacred' quality of the countryside which serves to exclude those who 'do not belong'. He illustrates how recent public order legislation has highlighted the status of non-belonging conferred on New Age Travellers, ravers and hunt saboteurs, each of which are seen to transgress the essential Englishness of the countryside.

In Chapter 13, Clare Fisher examines the complex interconnections between the rural imagination, and people living in the countryside and working in the craft industry. Here she notes that people are constructed as alternative, or 'other', yet at least in part they choose the position of other as demarcating a space which empowers their work. In so doing she offers very interesting glimpses of the interstices between more orthodox cuts of identity in rural studies.

Finally, in Chapter 14 Paul Cloke addresses questions about the otherness of the 'poor' and of 'poverty' in rural societies. Using illustrations from both rural England and the rural USA he demonstrates how discursive codes, symbols and concepts of poverty and welfare have differentiated spatially between the urban and the rural. Thus the political significance attributed to the issue of poverty is mediated by cultural constructions of rurality. However, there is a need (as perhaps in all of the subjects addressed in this book) for a more sensitive deconstruction of seemingly monolithic but actually chaotic 'othered' groups resulting from such constructions – in this case the rural poor.

REFERENCES

Bell, M. (1994) *Childerley: Nature and Morality in a Country Village*, University of Chicago Press, Chicago.

Bernstein, R. (1992) *The New Constellation: The Ethical–Political Horizons of Modernity/Postmodernity*, MIT Press, Cambridge, Massachusetts.

Bondi, L. and Domosh, M. (1991) 'Feminism and postmodernism: an uneasy alliance', *Praxis International* 11, 137–49.

—— (1992) 'Other figures in other places: on feminism, postmodernism and geography', *Environment and Planning D: Society and Space* 10, 199–214.

Cloke, P. (1994) '(En)culturing political economy: a life in the day of a "rural geographer" ', in P. Cloke, M. Doel, D. Matless, M. Phillips and N. Thrift, *Writing the Rural: Five Cultural Geographies*, Paul Chapman Publishing, London.

—— (1995) 'Rural poverty and the welfare state: a discursive transformation in Britain and the USA', *Environment and Planning A* 27, 1001–16.

—— (1996) 'Rural lifestyles: material opportunity, cultural experience, and how theory can undermine policy', *Economic Geography* 72, 433–49.

—— (1997) 'Rural studies: from country backwater to virtual village?', *Journal of Rural Studies* (forthcoming).

Cloke, P. and Goodwin, M. (1993) 'Rural change: structured coherence or unstructured incoherence?', *Terra* 105, 166–74.

Cloke, P., Milbourne, P. and Thomas, C. (1994) *Lifestyles in Rural England*, Rural Development Commission, London.

—— (1995) 'Poverty in the countryside: out of sight and out of mind', in C. Philo (ed.) *Off the Map: The Social Geography of Poverty in the UK*, Child Poverty Action Group, London.

—— (1997) 'Living lives in different ways? Deprivation, marginalisation and changing lifestyles in rural England', *Transactions of the Institute of British Geography* (forthcoming).

Cohen, A. (1994) *Self Consciousness: An Alternative Anthropology of Identity*, Routledge, London.

Daniels, S. (1993) *Fields of Vision*, Polity Press, Oxford.

Deleuze, G. (1990) *The Logic of Sense*, Columbia University Press, New York.

Doel, M. (1990) 'Deconstruction on the move: from libidinal economy to liminal materialism', *Environment and Planning A* 26, 1041–59.

Gregory, D. (1994) *Geographical Imaginations*, Blackwell, Oxford.

Halfacree, K. (1993) 'Locality and social representation: space, discourse and alternative definitions of the rural' *Journal of Rural Studies* 9, 1–15.

Haraway, D. (1990) 'Reading Buchi Emeta: contests for women's experience in Women's Studies', *Women: A Cultural Review* 1, 240–55.

Harvey, D. (1989) *The Condition of Postmodernity*, Blackwell, Oxford.

—— (1990) 'Between space and time: reflections on the geographical imagination', *Annals of the American Association of Geographers* 80, 418–34.

—— (1993) 'Class relations and social justice', in M. Keith and S. Pile (eds) *Place and the Politics of Identity*, Routledge, London.

Ignatieff, M. (1994) *The Needs of Strangers*, Chatto and Windus, London.

Jewitt, S. (1995) 'Europe's "Others"? Forestry policy and practices in colonial and postcolonial India', *Environment and Planning D: Society and Space* 13, 67–90.

Keith, M. and Pile, S. (1993) 'The politics of place', in M. Keith and S. Pile (eds) *Place and the Politics of Identity*, Routledge, London.

Lacan, J. (1977) *Ecrits*, Tavistock, London.

Lewis, B. (1993) *Islam and the West*, Oxford University Press, Oxford.

Marsden, T., Murdoch, J., Lowe, P., Munton, R. and Flynn, A. (1993) *Constructing the Countryside*, UCL Press, London.

Mead, G. (1934) *Mind, Self and Society*, University of Chicago Press, Chicago.

Morley, D. and Robbins, K. (1995) *Spaces of Identity*, Routledge, London.

Morris, M. (1992) 'The man in the mirror: David Harvey's "Condition of Postmodernity" ', *Theory, Culture and Society* 9, 253–79.

Murdoch, J. and Pratt, A. (1993) 'Rural studies: modernism, postmodernism and the "post-rural" ', *Journal of Rural Studies* 9, 411–28.

—— (1994) 'Rural studies of power and the power of rural studies: a reply to Philo', *Journal of Rural Studies* 10, 83–7.

Phillips, M. (1997) 'Rural restructuring: social perspectives', *Journal of Rural Studies*, (forthcoming)

Philo, C. (1992) 'Neglected rural geographies: a review', *Journal of Rural Studies* 8, 193–207.

—— (1993) 'Postmodern rural geography? A reply to Murdoch and Pratt', *Journal of Rural Studies* 9, 429–36.

Pile, S. and Thrift, N. (1995) 'Mapping the subject', in S. Pile and N. Thrift (eds) *Mapping the Subject*, Routledge, London.

Rocher, R. (1993) 'British Orientalism in the eighteenth century: the dialectics of knowledge and government', in C. Breckenridge and P. van der Veer (eds) *Orientalism and the Postcolonial Predicament*, University of Pennsylvania Press, Philadelphia.

Said, E. (1978) *Orientalism*, Pantheon, New York.

—— (1986) 'Orientalism reconsidered', in F. Barker, P. Hulme, M. Iverson and D. Loxley (eds.) *Europe and Its Others*, University of Essex, Colchester.

—— (1993) *Culture and Imperialism*, Chatto and Windus, London.

Schatzki, T. (1993) 'Theory at bay: Foucault, Lyotard and politics of the local', in J. P. Jones III, W. Natter and T. Schatzki (eds) *Postmodern Contentions*, Guilford, New York.

Shurmer-Smith, P. and Hannam, K. (1994) *Worlds of Desire, Realms of Power: A Cultural Geography*, Arnold, London.

Sibley, D. (1995) *Geographies of Exclusion: Society and Difference in the West*, Routledge, London.

Swyngedouw, E. (1995) Book review: 'Geographical Imaginations' by Derek Gregory, *Transactions of the Institute of British Geographers* NS 20, 387–9.

2

OF OTHER RURALS?

Chris Philo

BY WAY OF INTRODUCTION: A TRIP DOWN SHAKER LANE

Let me begin my chapter by discussing a children's story called *Shaker Lane* (Provensen and Provensen 1991), which tells of some changes that affected a small community living in and around Shaker Lane, an imaginary place located in the rural United States. According to the story, this place had first been settled by Shakers, members of an unusual religious sect that I will mention again shortly, and we are told that a Shaker Meeting House had once stood at the crossroads of Shaker Lane and School House Road. The Shakers had long departed this place, though, and in their wake a new and in some respects equally curious collection of people had arrived to settle and to shape a new rural landscape. In the big house on the hill lived two elderly women, Abigail and Priscilla Herkheimer, who owned much of the local land but were having to sell it piece by piece; and gradually, as the lots were sold, a row of houses had grown up along the lane supporting a tiny community. There was Virgil Oates and his wife, Sue Ann, their five children and Sue Ann's brother, Wayne; and there was Sam Kulick next door, with Norbert Le Rose and his wife and children and animals across the road; and there was Old Man Van Sloop with his dogs, goats, chickens and bottles of this and that; and there were the Whipple twins who did everyone's gardening; and there was the Peach place, with Big Jake Van der Loon who could do anything (he dug wells, moved barns, put up fences); and then there were quite a few other folk besides.

This new Shaker Lane community stood a little out of the 'mainstream' – its residents being fairly marginal kinds of people, poor but relatively self-sufficient, friendly but insular, wary of the outside world, holding values about family and community life but not being materialistic – and at the same time their imprint upon the local environment was one which conventional society would probably find unacceptable:

The people who lived on Shaker Lane took things easy.
Their yards were full of stuff –
old dressers waiting to go inside,
carts that would never roll again,
parts of old trucks,
stove-pipes, piles of rotten wire and tin cans.
Some people would have liked to see Shaker Lane disappear forever.
When the big yellow school bus came down Shaker Lane,
the kids would yell'
'Aker, baker, poorhouse shaker!'
Sometimes there were fights.

(Provensen and Provensen 1991: no pagination)

The Shaker Lane folks lived in, and contributed to, a messy rural landscape: one that was full of unlike objects and land uses all mixed together, with badly painted shacks and weed-infested vegetable gardens blurring into one another, with an absence of fences and delimited plots, with old furniture, rusting cars, rubber tyres, empty bottles, dogs, cats and chickens chaotically jumbled around the central lane (see Figure 2.1). This was certainly not a picturesque village, a far cry from either the manicured or the wilderness rural scenes of many mainstream imaginings, and as such it possessed a socio-spatial order that was very much its own rather than reflecting conventional assumptions about how society and its spatial supports should be organised.

One day the Shaker Laners received a visit from Ed Rickert, the County Land Agent, and he brought the news that a reservoir was to be constructed in the vicinity, and that the County was going to place a compulsory purchase order on the land presently occupied by the residents of the Lane. These folks could hardly fight the order – they were not people with any political clout, money for lawyers, or even basic knowledge of their rights – and so, with resignation, sadness and not a little anger, they packed up their belongings, loaded them onto their clapped-out cars and vans, and moved out. As they did so the bulldozers moved in, digging out a vast hole where once had passed much of the Lane; and then the waters of the reservoir began to rise, covering the houses that remained, except for the house on the hill where the Herkheimer sisters still lived. A year or so later what was left of Shaker Lane had its name changed to Reservoir Road, and from then on it quickly acquired a set of residents rather different from those who had been displaced. As the Provensens put it, 'you would not know the place' (Provensen and Provensen 1991: no pagination), and the ramshackle geography of the old Lane was now turned into an estate of large houses, smart and immaculately kept with neat gardens full of swimming pools, basketball courts and afternoon teas being taken on tidy plastic garden furniture. The chaotic spaces had been compre-

Figure 2.1 Shaker Lane
Source: from Shaker Lane © 1987 Alice and Martin Provensen. Reproduced with permission of Walker Books Ltd.

hensively cleaned up, and it could be argued that an entirely new socio-spatial order – one reflecting values about order, private property, material possessions – had now become inscribed into this particular piece of the rural landscape. The result was hence a classic case of what David Sibley might call the 'purification of space' (Sibley 1988, 1995: especially chs 4–6), which so often occurs when the apparently disordered socio-spatial worlds of 'outsiders' such as Gypsies, 'Travellers', nomads and indigenous peoples are steamrollered away, unwittingly or otherwise, by the powerful forces of Western society hung up on a rigid – geometrical, hierarchical, logical – sense of order.[1]

WAITING FOR THE CALL OF 'THE OTHER'?

The purpose of this strange introduction has been twofold. On the one hand I wanted to introduce in an immediate and substantive sense the possibility of taking seriously rather different rural peoples than are commonly considered in the academic literature: the possibility of looking into areas that get called 'rural', and of recognising that living and maybe working in such areas there will be numerous different human groupings, many of which will depart some way from the usually valued norms of Western society, and many of which will also possess relationships with space and place which depart from what is commonly regarded as normal. The humble folks of Shaker Lane, living in their ramshackle houses amid chaos and junk, fall into such a category, and hence can be understood to comprise what I am calling – albeit with some trepidation – a particular 'rural other' whose imprint on their local surroundings creates what might be called an 'other rural' (although the basis for this identification of what is 'other' is itself highly problematic and will be returned to in conclusion). The principle of taking seriously this coming together of 'rural others' and 'other rurals' underlay an earlier essay of mine (Philo 1992), where I urged academic rural geographers to enlarge the horizons of their studies by wondering about the worlds of many more non-hegemonic, commonly less than visible, often sad and oppressed, sometimes defiant and resourceful 'rural others' than have to date been touched upon. I certainly did not mean to imply that rural researchers have never considered such peoples, and I am sure that numerous anthropological and sociological inquiries of relevance in this respect can be found, and neither was I claiming that everything to date conducted by rural geographers is inescapably flawed by a neglect of otherness.[2] In addition, it is evident from various sources – this book included, but also, for instance, recent issues of the *Journal of Rural Studies* (e.g. Bell and Valentine 1995; Brandth 1995; Davey and Kearns 1994; Mackenzie 1994) – that there is presently an exciting swirl of interest within and around rural geography about the conceptual, empirical and perhaps 'political' implications of paying attention to different kinds of people occupying, experiencing, shaping and coping with different kinds of spaces.

All of this being said – and this is the second purpose of my introduction – I would want to argue that a tendency still exists in and around the rural geography literature (as it does in many other literatures) for academics to proceed in a fashion analogous to how the planners, developers and Reservoir Road inhabitants imposed a strict order upon the messiness of the Shaker Lane community. Indeed, there remains here a risk of continually replicating the thoughts constitutive of what Michel Foucault terms 'the Same': 'the order imposed on things', the conceptual processes by which phenomena are 'distinguished by kinds and ... collected together into identities', the largely taken-for-granted patterns through which given societies name, envisage, separate and relate the things (material and immaterial) of the world (see especially Foucault 1970: xxiv). Being locked into the thought-prison of 'the Same' means that it becomes impossible to appreciate the workings of what Foucault then terms 'the Other': 'that which, for a given culture, is at once interior and foreign, therefore to be excluded (so as to exorcise the interior danger) but by being shut away (in order to reduce its otherness)' (Foucault 1970: xxiv). Alan Megill suggests that 'Foucault sees a conflict in history between "the Same" and "the Other"' (Megill 1985: 192; see also Philo 1986), and it can readily be appreciated how various of Foucault's historical studies from *Madness and Civilization* (Foucault 1967) onwards have displayed a concern for the actual power struggles arising as Western society's guardians of 'the Same' strive to control the embodied representatives of 'the Other'. However, elsewhere in his corpus (e.g. Foucault 1970, 1972) the focus has been much more on the devices of thought, language and discourse which serve to spread the hegemony of 'the Same' (at bottom, Western reason) while fixing, defusing and reducing to silence and nothingness all irruptions of 'the Other'.[3] Foucault would claim that *all* Western intellectual endeavours are implicated in this action of conceptual oppression, even if there is scope for resistance, and I would thereby acknowledge that the specific corner of the intellectual division of labour known as 'rural geography' must itself be so implicated as well.

It is possible to enlarge this analysis through the claims of Marcus Doel, who draws upon the demanding texts of Jacques Derrida, Gilles Deleuze and others to illustrate how intellectual exercises conventionally carry with them an impulse to deal in terms of 'the Other of the Same' as opposed to approaching the possibilities of 'the Other of the Other' (see especially Doel 1994a; and also Doel 1993, 1994b). More specifically, what Doel argues is that in many studies, and despite common protestations about wanting to get close to different people, places, practices at the same time as remaining open to new categories, connections, concepts, the result all too often fails to do more than simply reaffirm established modes of thought (more or less inadvertently obscuring differences by revisiting familiar terms, comparisons, theories). This is not just because it is difficult to conduct the innovative research projects that are needed in the field – although methodological issues should not be overlooked – but, and as Doel insists, it is because the very equipment for thought bequeathed to scholars in the West has an inherent 'will' to translate what might initially appear as somehow 'the Other' into the comforting vocabularies of what is tried and trusted as 'the Same'. Here is how Doel rehearses the distinction between 'the Other of the Same' and 'the Other of the Other':

> While the former belongs to the theoretical-practice, falling under its jurisdiction and influence, the latter does not. More precisely, in the former difference and otherness are *appropriated* to the Same, while in the latter they are *ex-appropriated*. Writing about otherness can therefore follow one of two trajectories . . . *to the right*: appropriation, overcoding, territorialisation, accumulation and capture; *to the left*: ex-appropriation, decoding, deterritorialisation, expenditure and flight.
>
> (Doel 1994a: 1042)

These are complex notions, but in the context of rural geography I think it evident that the subdiscipline has tended to operate on the terrain of 'the Other of the Same', keeping its subject-matters firmly anchored in familiar empirical and conceptual moorings. In so doing it has effectively simplified the countryside – whether by being hung up on agriculture (by tracing all forms of social relations back to the farms and fields); by a fascination with the neat morphological unit of the nucleated village; by an obsession with *Gemeinschaft* social relations; by a persistent questioning of the local–newcomer schism as a key division (even if seen as cross-cut by class divisions: see Cloke and Thrift 1990) – with the upshot that studies in rural geography have an element of predictability about them. I will qualify this statement later, but to an extent readers do tend to encounter again and again the 'same rurals' (notably either the picturesque village or an uplands wilderness) populated by 'rural sames' (country squires, yeoman farmers, landless labourers, gentrifiers and geriatrifiers, most of whom are conceived of as white, middle-aged men),[4] and as such the possibilities that a more inclusive rural geography might touch are severely restricted. Moreover, the conceptual operations of rural researchers – whether they are mapping out 'central place systems', theorising the workings of the 'rural state', recovering 'senses of place' from village-dwellers, or whatever – can often serve to overlay a curious conceptual sameness on the specific rural contexts encountered in the field, however unusual or individual these may initially have seemed. There is a capturing of the many possible rural worlds, a development that amounts either to ignoring possibilities (they are deemed irrelevant to the forces of 'real' rural change) or to their translation into alien frameworks (which can involve 'territorialising' or fixing them within a pre-determined and not all that imaginative account of the countryside). Such a tendency leads directly to a neglect of those 'other rurals' like Shaker Lane which appear to have little relevance to the overall trajectory of rural change, except occasionally as its causalities, or which stand some way outside of the usual frameworks through which the rural scene is conceptualised.

Doel's reaction to this state of constantly rewriting 'the Other of the Same' is to wonder about tactics which will enable academics to gain a better glimpse of 'the Other of the Other'. The ambition is to let otherness somehow arise in intellectual work upon the world, to permit 'the Other' (in the guise of things different, unexpected, unfathomable) a hearing at the same time as room is made for an encounter with the circumstances and the voices of all manner of specific 'others' who have previously been marginal to the academic gaze (and who have also not participated at all extensively to date in the hallways of the academy).[5] Doel discusses four different strategies that might be employed in this respect, each of which amounts less to a method of how to proceed and more to a 'philosophical' stance which can be taken before substantive materials, but the prevailing image that he conveys is one of 'telephony': of needing to wait for the call of 'the Other', and in the process inevitably awaiting calls from quite specific 'others' struggling to get by in their everyday lives. He thereby talks about trying to ensure that 'the Same (and its Other) is withdrawn into the open multiplicity of a telephonic switchboard which is always on call/hold for the interruption of an-Other, and an-Other, and an-Other, and . . .' (Doel 1994a: 1048). Furthermore, he insists that this is a stance which is promised, if not demanded, by some of the most recent intellectual currents which have affected human geographers:

> postmodernism, poststructuralism and deconstruction proceed through an affirmation of the otherness of the Other. Specifically, they work through the ethical obligation to respond to the call of an-Other. . . . 'Interrupt me', requests Derrida. In ex-appropriating itself to an-Other who is *always to come*, the telephonic switchboard short-circuits the apparatuses of capture which try to *force* an-Other to become the Other of the Same. . . . It is in this sense that writing difference is both interminable and impossible: in placing itself on call/hold for an-Other, theoretical-practice is obligated to wait for an-Other whose arrival is therefore always *to come*.
>
> (Doel 1994a: 1047)

In the spirit of what Doel is advocating here, if rural geographers are to begin critiquing processes of power whereby different rurals become marginalised – and if in so doing they are to worry about how these different socio-spatial worlds might be represented – then 'we' should remain (metaphorically speaking) by our 'telephones', alert to the often surprising things that we might be told by 'rural others' calling long-distance from 'other rurals' that are apart from us, alien to the intellectually competitive academy where most of us work, distinct from the sanitised, secularised, materialistic social spaces from which most of us hail and into which most of us have been socialised. And in the next section of this chapter let me strive to do just this, although doubtless I will fail to throw off all (or even many) of the 'apparatuses of capture' demanded by 'the Same': and let me turn my attention to what strikes me at least as a quite 'other rural'.

ATTENDING TO AN 'OTHER RURAL': SHAKERS IN THE COUNTRYSIDE

In this section let me return to Shaker Lane, but this time I want to say something about the original community that is presented as having first settled this particular part of the rural United States: namely, the so-called 'Shakers' themselves, whom in the late eighteenth and early nineteenth centuries set out to create new religiously charged spaces different from the secular society and settlement found elsewhere in the Western world. My discussion here can only scratch the surface of matters, and is admittedly based on a synthesis of secondary materials rather than on my own primary research,[6] but I hope that what follows will nevertheless spotlight an 'other rural' which is in itself quite remarkable and worthy of geographical study.

Shaker theology, history, geography

The Shakers were a religious sect or band who appeared in England (around Manchester) during the mid-eighteenth century, and who quickly shifted their operations to North America as one group of pioneering settlers among many others (a number of which had a religious orientation deeply marking the character, history and geography of their settlement).[7] The Shakers – or the 'United Society of Believers', as they were also known – were 'millenarianists' in that they believed the second coming of Christ would be followed by a thousand years of 'heaven on earth' prior to the final judgement. To be more precise, they were 'pre-millenarianists' because they supposed Christ already to have arrived on earth and to have revealed himself to Mother Ann Lee, an individual who was foundational to the whole Shaker theology. The notion that heaven and earth were now bound together, and that a peaceful and godly pattern to life was now attainable (and indeed should be striven for) on earth, was central to how the Shakers conducted themselves in their everyday affairs. Indeed, they subscribed to a doctrine of 'perfectionism' which meant that they were continually seeking to achieve heavenly perfection in their practical activities on earth, such that in all that they did – from the largest of tasks (building a church, cultivating a wilderness) to the smallest (making a wheelbarrow, weaving a basket) – they were both making heaven concrete and glorying the God who was the source of all perfection. This was a theology obsessed with order, moulded by the vision of an external secular world totally disordered after the fall of Adam and Eve, and convinced that it was only in a few havens of spiritual enlightenment that the message of Christ's second coming could be heeded in a reordering of human society commensurate with the peace, perfection and order of heaven. What is particularly interesting is the way in which this heavenly order was envisaged by the Shakers, given that there was a double movement occurring whereby the existing patterns of Shaker life on earth were projected onto heaven – creating a Shakerised 'imaginary geography' of heaven – at the same time as this vision of God's province was taken as a blueprint for the organisation of Shaker life in the here and now. Dianne Sasson sums up this double movement when examining an image of a building in heaven, a temple not dazzling but 'plain and simple', which was described in the spiritual narrative of an individual Shaker:

The building in Green's vision reveals the complexity of Shaker metaphors, for it points in two directions at the same time. It is an image
of the earthly community of Believers and reflects Shakers' belief in
simplicity, utility and lack of ornamentation. At the same time, it
suggests an ideal spiritual world, a divine achetype which the earthly
community seeks to emulate.

(Sasson 1983: 41)

The significance of these points about 'heaven on earth', the pursuit of
order and the doubling of heavenly images with earthly realities will be
developed further on.

As indicated, a crucial figure for the Shakers was a woman called Ann
Lee who claimed to be a 'female Christ' on the basis of a 'grand vision'
that she experienced in 1770, the essence of which involved Christ 'dwelling
within her' as a husband, a lover, an equal. Shaker theology hence embraced
the controversial idea of a dual manliness and womanliness residing in the
entity of a 'bisexual god' (Cross 1950: 31) – what one writer terms
the 'maternal principle in deity' (Patterson 1979: 19) – and this idea translated into a dual leadership of daily Shaker affairs whereby for every man
in a position of responsibility there would also be a woman. Unsurprisingly, such arrangements have attracted claims about the Shakers as
exponents of a feminist theology, it being described as an 'ascetic feminism'
(see Desroche 1971: especially 77–9 and ch. 5) and as feeding into 'doctrines
impossible of execution by unregenerate man' (Cross 1950: 31). Moreover,
Mother Ann preached that a life of total celibacy was the only way in
which human beings could escape from the depravity and despair into
which earthly human society had now fallen, implying that Shakers had
to commit themselves to a life of complete celibacy – they were to sustain
themselves as a sect by continually recruiting new members, if possible
ones with children, rather than through their own natural increase – and
in so doing they had to abolish the conventional family unit. Man
and woman could not live together as husband and wife within a Shaker
community, and new recruits who were conventionally married were
expected (if successfully initiated) to go through a service of *de*-marrying.
Instead of living in family units, and instead of occupying neatly set apart
family or private spaces, men and women lived together in much larger
communal units (these units constituted the Shaker 'families') and spent
virtually all of their time in common public spaces where work, rest, play
and worship were shared. At the same time men and women were kept
very much separate from one another – men's space and women's space
remained quite distinct even though the public–private divide was largely
broken down[8] – and this meant that the micro-spaces of Shaker settlements
were tightly regimented and bound by formal rules of who could go where
(a theme that I will return to). I have laboured this issue of the Shaker

27

approach to gender and sexuality because I immediately wish to signal how unusual, 'other' and even shocking certain aspects of Shaker life were (and still are) to mainstream religious, gender and sexual sensibilities.

Ann Lee's vision in a Manchester gaol in 1770 is sometimes taken as the starting-point for the Shaker sect, but it had in fact emerged a little earlier in the guise of a grouping which had splintered off from the Quakers of Bolton in the 1740s and subsequently moved to Manchester. This grouping soon became known by the name of 'Shaking Quakers' or 'Shakers' thanks to the remarkably frenzied dances which occurred at their meetings, bodily movements which were later to be subjected to a strict discipline, and it was this pre-existing religious milieu which provided Ann Lee with the conceptual and practical resources to press home the implications of her visitation. Another vision, as witnessed by one of her associates called James Whittaker, persuaded her that the Shakers needed to travel to America to establish themselves in a more organised fashion, and in 1774 Mother Ann and a small band of followers duly crossed the Atlantic. In 1776 these early Shakers were able to assemble as a settled community at a small place called Watervliet, near Niskeyuna in New York state,[9] itself an intriguing place to have chosen because it was on the fringes of a region which, under the name of the 'Burned-Over District', was later to become well known for its concentration of 'people extraordinarily given to unusual religious beliefs' (Cross 1950: 3). A handful of dedicated Shaker missionaries (Mother Ann herself included) then operated out of Niske-yuna during the late 1770s and 1780s, apparently attracting hundreds of converts, and, faced with the task of setting up earthly arrangements for these many converts, the sect developed the strategy of founding 'specialist' settlements in which whole Shaker communities could exist free from outside interference. This strategy was very much the idea of Father Joseph Meacham, who took over the leadership of the sect after the deaths of Mother Ann and Father James, and under whom a directive instructing members 'to join together' ensured that 'geographical consolidation replaced itinerant evangelism' (Stein 1992: 43). This process was begun with the establishment in 1787 of a second settlement at a site called New Lebanon in the east of New York state, not far from the border with Massachusetts, and it was not long before a number of other Shaker settlements began to appear.

The Shakers were not alone in developing such a 'communitarian' approach – some have even called it a kind of 'communism'[10] – but the evidence suggests that it was they rather than a number of other companion religious and secular groupings who were most accomplished in the creation of new spaces and places where their distinctive beliefs could be translated into a cohesive mode of everyday living:

Between 1780 and 1826 the Shakers founded twenty-five settlements from Maine to the Ohio frontier. These well-built and prosperous villages greatly enhanced the credibility of the communitarian strategy for social change in the United States. The existence of over two dozen model Shaker communities was cited by communards everywhere as proof that it was possible to construct a satisfactory model community and then to duplicate such an environment on demand.

(Hayden 1976: 65)

An overall Shaker geography began to emerge, entailing a series of locations spread across the eastern seaboard – ones that were some way behind the so-called 'frontier'[11] – and with a few 'colonies' and 'mission stations' situated further west whose occupants were very much caught up in the frontier experience.[12] There were obviously all manner of peculiar local factors bound up in the precise locations taken by settlements, in part to do with the distribution of affordable and bequeathed land, and it is also clear that some sophisticated agricultural decision-making was displayed by individual Shaker communities when 'gather[ing] on the best lands available to them' (Sprigg and Larkin 1988: 38–9). In part too, though, practical considerations blended with the more intangible spiritual influences that impelled individual Shaker communities to choose certain sites rather than others for either whole settlements or specific buildings and fields (as in the case of one Mother Lucy who 'felt a gift' to have a house put up at a site near the carding mill of the New Lebanon settlement). The inspired 'rightness' of Shaker locational decisions has sometimes been commented on, notably in an auto-ethnographic piece by John McGuire where he writes of beginning 'to understand the Shakers' choice of Hancock for their village', and reflecting on both 'the serenity of the setting' and 'its obvious commercial potential, being located near transportation routes' (McGuire 1989: 35).

A number of individual Shaker settlements were established, then, perhaps supporting a maximum of 6,000 people at the height of the movement in the 1840s (although detailed work on the 1840 Census reveals a total of only 3,608 members). Many of the individual settlements were located at sites quite distant from one another, and a feel for the difficulties faced by Shakers seeking to move between settlements is captured in a song telling of a visit made by three elders to Busro in Indiana during a period of heavy flooding, a trip of 'near seventy miles' through a 'howling wilderness ... a thousand furlongs in extent' (in Patterson 1979: 142–4). Despite such difficulties of distance and terrain, the Shakers remained a tightly integrated sect with strong connections binding the isolated communities to the 'capital' of the Shaker world, New Lebanon, from which its senior members sought to control the operations of the sect throughout North America. Indeed, the aim was to ensure a strict conformity in Shaker practices from one Shaker village to the next:

Each of the Shaker villages was like the others, not only in the organisation of its religious and temporal affairs, but in its architecture, in its customs and folkways, in its dress and the speech of its inhabitants, and in the general nature of its agricultural, horticultural and industrial art activities. Such similarities were brought about and augmented by the rigid regulations of the central ministry, the periodical visitation of this ministry to the several [communities], the co-operation of different communities in constructional enterprise, the frequent movement of members from one village to another, and the interchange of goods.

(Andrews and Andrews 1982: 26)

It follows from the great emphasis which the Shakers placed upon order that there should be this deliberate attempt to bind all of the component communities and members into a firm structure, one seeking to impose conformity through similarity and hence battling against differences arising for whatever reason (including those rooted in geography). At the heart of this structure were the 'Millennial Laws' issued by Father Joseph in 1821 and then reformulated in 1845. One commentator describes the later code as regulating 'everything imaginable' so as 'to guarantee identical practice from one community to the next' and hence to create the above-mentioned 'perfect reflection of the heavenly in the earthly' (Whitson 1983: 20). In reality, there were of course lapses from conformity and similarity, and Priscilla Brewer argues that 'controlling these small societies from New Lebanon was difficult', citing the case of the Savoy settlement in Massachusetts where 'Believers ... were free from the more severe strictures that would have governed their lives at more solidly established Shaker villages' (Brewer 1986: 38). Moreover, it appears that there was some debate within the Shaker leadership about the dangers of over-regulating the movement, and the 1845 code was actually withdrawn after five years because it was felt that the imposing of too much uniformity could compromise the authenticity of how different communities and members experienced their oneness with God. Even so, there can be no doubting that the Shaker world was a thoroughly organised one with a high degree of socio-spatial coherence bolstered by an 'extensive communications network' (Stein 1992: 66), and that they differed in this respect from a grouping such as the Amish whose organisation 'lack[ed] a geographical focus and a directive-issuing church hierarchy' (Crowley 1978: 263).

Ruralising settlements, activities, moralities

The emerging geography of the Shakers was very much a rural one, in part because that was where land was readily available and in part because the wide open spaces offered the possibility of 'retreating' from wider society. In this latter respect it is evident that the 'utopianism' of the Shakers, their wish to build utopias of heavenly order on earth, fuelled a drive to separate themselves from the disordered and contaminating ways of ordinary people living, working and procreating in ordinary places. Revealingly, the term that they used to refer to these people – 'world's people' – was one which quickly acquired the dual meaning of 'people outside [of] the faith and physically outside of the geographic boundaries of the [Shaker] community' (McGuire 1989: 24), thus underlining the extent to which Shakers conceived of themselves as necessarily closed off socially and spatially from the everyday world and its occupants. It appears that Father Joseph was pivotal to this development:

> [He] conceived the call to perfection in Christlife as a call to separa-
> tion: separate the regenerate–resurrectional way of life from that of
> the world. Henceforth, he saw the ideal as Believers living completely
> apart from the 'world's people', as Believers tangibly developed a
> perfect life in which the practical things of living would directly flesh
> out the inner Christ spirit. The new communes would enable Shakers
> to practice celibacy without outside pressures; they could share the
> things of life in a complete religious communism; they could live at
> peace without violence or involvement with those who were violent.
> (Whitson 1983: 16)

On religious grounds the Shakers became increasingly perturbed by an external society which they viewed as an 'antechamber to perdition', and so they endeavoured 'to insulate themselves from its contamination' (Whitworth 1975: 28), but it also appears that on a variety of more secular grounds the entry to 'withdrawn settlements' was designed to prevent 'entangling alliances' developing between individual Believers and old friends, their 'natural' families and even organisations such as political movements (a point underlined by Melcher 1941: 51). It might be argued that the 'isolationist policy' of the Shakers has been overstated, and it should be acknowledged that in an economic sense Shaker villages were commonly bound into worldly trade networks, and that their products – furniture, baskets, medicinal herbs, garden seeds, dried fruit and vegetables – were actually sold far and wide to meet a considerable demand for quality items (as is stressed by Andrews and Andrews 1982: 40–1; Rubin 1984; Sprigg 1986; Sprigg and Larkin 1988).[13] Robley Edward Whitson, meanwhile, highlights the facts that many Believers have 'lived outside the communal societies' and that 'the earliest Shakers, including Mother Ann, did not envision forming communes' (Whitson 1983: 28), although these points surface in an idiosyncratic reading of Shaker theology as being less separatist and more worldly than is usually acknowledged. It is none the less instructive to return to the views of Mother Ann, since it is known that she definitely did react adversely to 'the racket and filth of the Man-chester slums where she was raised' (Sprigg 1986: 16–17). In so doing she managed to prefigure two prominent dimensions to subsequent Shaker thinking, the first of which was an obsession with cleanliness that took Ann's declaration of '[t]here [being] not dirt in heaven' (in Sprigg 1986: 17) as a mandate for creating sanitised Shaker places on earth sealed off from any inward transfer of the world's dirt.

The second dimension here – and one that indicates the rural to have been much more than just an accidental site for Shaker activities – was an acute anti-urbanism rooted in how Shakers were seeking to distance themselves, not just from ordinary society, but more specifically from the hustle of the urban arena. As one writer explains, 'their one desire, when-ever they were obliged to go on business to "great and wicked cities", was to get back as soon as possible to their homes in the country' (Melcher 1941: 120), and this also meant that a key locational consideration for their settlements was 'a sufficient degree of remoteness from towns to ensure the isolation [that] their owners desired' (Melcher 1941: 122). Towns and cities were viewed as the prime earthly environments in which the sins of humanity (particularly the satisfaction of sexual desire) were concentrated and repeatedly acted out, and so the Shaker alternative was unsurprisingly to seek out instead a place in the countryside where virtue rather than vice could flourish. Various Shaker documents testify to this rejection of the urban in favour of the rural, most notably a Shaker hymn which sings of living in a hamlet 'remote from the thronged thoroughfares where business and pleasure with folly abound', while also celebrating being able in this hamlet to 'dwell in retirement and breathe the pure airs' (in Hayden 1976: 75). The theme of the countryside as a source of uplifting beauty, purity and spiritual refreshment recurs in other Shaker songs, and specific ones tell of finding a 'healing balm' in the 'verdant valley', of 'one pleasant grove' restored to a glory 'where the fig tree is forever bearing', and of the 'lovely objects of my love' which are 'my flock, my vineyard, my dove' owned in common with 'one little band' (the Shakers) preferred by the singer out '[o]f all the sects that fill the land' (in Patterson 1979: various pages).[14] The autobiographies of individual Shakers also contain positive images of the rural, sometimes in the context of presenting the Shakers as enabling 'God's restoration of a paradisiacal garden in the wilderness of America' (Sasson 1983: 32–3), but often as part of a structure familiar to spiritual narratives which speaks of arduous travels through rugged rural landscapes – wildernesses, swamps, mountains and rivers – being rewarded by reaching a holy destination. The deep morality of a rural existence was a common perception of many late eighteenth-century religious groups, but for the Shakers this connection was felt particularly strongly, in part because the hard manual labour demanded by working in the fields was considered a powerful 'antidote to lust'. This meant that agriculture and a countryside setting were considered crucial to the successful running of a Shaker community,[15] and one writer summarises what was involved here by indicating that 'Shakerism was grounded in the traditions of agriculture and handicrafts, and was close to the lore of the countryside. Agriculture was primary: "[e]very commune, to prosper, must be founded, so far as its industry goes, on agriculture. Only the simple labours and manners of a farming people can hold a community together," declared Frederick Evans [a well-known Shaker]' (Harrison 1979: 174).

The relationship of the Shakers to the urban arena must be explored further, however, given that cities could on occasion be given a more positive gloss:

> settlements were also described in the religious tradition of the Sermon on the Mount: 'Ye are the light of the world. A city that is set upon a hill cannot be hid.' The Shakers dealt with the conflict between city and country very neatly, by giving each of their settlements two names, one corresponding to the rural village where it was located, and the other a 'spiritual name' suggestive of the Heavenly Jerusalem, such as 'City of Peace' or 'City of Love'.
>
> (Hayden 1976: 15)

The implication is that any one Shaker community did not merely inhabit an earthly location securely embedded in rural seclusion, but also was imagined at a heavenly location set in a large and beautiful city proclaiming the lasting glory of God. Surviving Shaker artwork hence includes representations of earthly dwellings 'becoming' heavenly buildings made of precious stones and containing fabulous objects, while two Shaker songs respectively mix up rural images (vines springing from desert ground) with urban ones (the hill-top city radiating a 'dazzling light') and cast Union Village, Ohio, as a 'beautiful city' criss-crossed by streets paved with gold (in Patterson 1979: various pages). This Shaker sense of the rural as simultaneously urban – albeit when the latter was conceptualised as existing in a different (heavenly) order of reality – suggests the envisioning of a quite 'other rural' to that possessed by non-Shakers, who it seems often became confused in reporting that small Shaker communities were sited next to magnificent cities strangely unnoticed on maps of the time. But one further twist still remains to be noticed in this connection, since numerous Shaker depictions of heavenly cities actually cast them as somewhat humble, downbeat affairs looking very like existing Shaker villages and certainly bereft of treasures: thus, 'the streets of the New Zion are not paved with gold but are straight and narrow paths through the wilderness; the buildings are not bejewelled palaces but the plain white buildings of the Shaker community' (Sasson 1983: 41). The humility of the Shaker mind-set came to the fore here, in that it was presumed by many Shakers that a truly godly realm would be free from the superficial bounty and grandeur hankered after by earthly people and hoarded in their degenerate cities. It may therefore be appropriate to identify two parallel conceptual mappings which were shaping Shaker thought in this respect, one from the rural to the urban and back again and the other from the earthly to the heavenly and back again, and the upshot was a complex amalgam of 'imaginary geographies' framing the Shakers' retreat to the countryside.

Regulating nature, space, bodies

It is important now to consider the practices of the Shakers within and around their rural communes, and in so doing to keep in mind the longing for order which permeated the whole sect and led to detailed prescripions being laid down for everything 'from the colour of the buildings to the placing of the right leg into a pair of trousers first' (McGuire 1989: 24). The initial thing to notice is that, leading from what was claimed above about the mobilisation of the rural–agricultural scene for moral ends, the root practices of the Shakers in relation to the natural world were designed to achieve a strict regulation and revalorisation of nature. Although the natural environment could be regarded as a spiritual resource, in other regards it was viewed with suspicion as the home of disordered and unrestrained passions released by the fall of Adam and Eve:

> In the physical landscape and in the creatures that inhabit it, Shakers saw an image of the wild and disorderly nature of unredeemed man [*sic*] who selfishly pursued sexual gratification with no regard even for the 'times and seasons' of procreation. The Shakers firmly believed that only a community based on divinely ordained laws could order and control nature in all its manifestations.
>
> (Sasson 1983: 38)

The implication was that the Shakers should devote their energies to (re)establishing a strict 'order and control' in relation to nature, the hope being to forge an island of heavenly order (the new 'Eden') safe from the surrounding seas of earthly disorder (fallen nature in its many guises). Seen in this light, it becomes clearer why the Shakers put such store by honing their skills as farmers able to exert complete mastery over brute nature, and this is why to them the cold earth was 'something to be redeemed from rugged barrenness into smiling fertility and beauty' (Andrews and Andrews 1982: 48).[16] This approach to agriculture as 'religious ritual' meant that the agricultural landscape produced by the Shakers was highly distinctive, being much neater, better-ordered and obviously productive than that of other farmers in the same neighbourhoods, and it was evident that 'Shaker farmers were not the same as other farmers (they worked more methodically, and to greater effect)' (Harrison 1979: 174). The superior agricultural techniques of the Shakers went hand in glove with substantial stone-wall fences, well-built barns and granaries, and creameries which were models of good design. What is more, the Shaker codes insisted that 'when brethren are about to farm, and find gates open, bars down or fences broken down, they should put them in order' (in Andrews and Andrews 1982: 52). Disorderly nature, sprouting weeds, sprawling hedges, straying animals and the like had no place in this ordered agricultural geography.

The same principles were of course repeated in the Shaker settlements themselves, and – unlike the later Shaker Laners of the children's story, who appear entirely disordered in a conventional sense – the Shaker villages were models of socio-spatial orderliness. There was a definite visual aspect to this orderliness, but there was also an underlying logic to how the society of each Shaker community was distributed across the settlement space in accordance with Father Joseph's conception of 'the communities as sanctuaries, using the imagery of the ancient temple to symbolise the new relationships' (Whitson 1983: 16). Outside of the temple walls were the 'world's people', whereas inside there were three 'courts': an innermost one or 'holy place' with the long-term committed Believers (the 'church order') and two other 'courts' housing respectively the newest recruits being introduced to Shaker ways (the 'gathering order') and those who were more experienced but still learning for their final consecration (the 'novitiate order'). This conception was cleverly translated into the social geography of individual Shaker communities, as organised around the so-called 'families' (sizeable collectives of men, women and children):

> Life in a Shaker village centred on the family. Each community consisted of two or three, and the larger ones had five or more.... Families were distinct entities akin to neighbourhoods and were located about a quarter to half-a-mile apart. Each family had its own dwelling house, workshop and barns. The central family was the Church Family (sometimes called the Centre Family). The meeting-house for the entire community was located there. Other families were named according to their geographic relation to the Church Family – East, North, West or South. Sometimes a family took its name for other reasons: Mill, Brickyard, Hill, Office.... Families also represented different orders, or levels, of commitment to the Shaker faith.
>
> (Sprigg 1986: 14)

If the broad social structure of each Shaker village was ordered according to strict principles, and in the process given spatial direction, much the same can be said about the routine daily social life of the various settlements. The 'Millennial Laws' decreed that there should be 'no talking, laughing, sneering, winking, blinking, hanging and lounging on the railings, hugging, fumbling and fawning over each other, when going to the table' (in Sprigg and Larkin 1988: 43), and numerous other rules barred members from wearing 'ragged clothes', failing to wash properly, using flowers for purposes of personal adornment, and the like. Interestingly, an intimate link was perceived between social order and physical order, as in the insistence that chaos could be avoided in dwellings and workshops only if 'furnishings and tools were... marked to indicate their proper location' (Sprigg and Larkin 1988: 44), and a preoccupation dating back to Mother Ann was the need to put things away in their proper place after use so that they would always be ready to hand.

To return to what June Sprigg calls 'a visual environment of... quiet power', it is obvious that the Shakers endeavoured 'to create a visible world in harmony with their inner life: simple, excellent, stripped of vanity and excess', and that in their efforts here 'the line between heaven and earth flickered and danced' (Sprigg 1986: 11, 21). It is thus unsurprising that '[c]ontemporary observers remarked [how] a Shaker village looked different from an ordinary village (it was neater, cleaner, better planned)' (Harrison 1979: 174); nor that '[i]n 1851 Nathaniel Hawthorne found everything so neat at Hancock village, Massachusetts, that he said it was "a pain and constraint to look at it"' (Sprigg and Larkin 1988: 176); nor that a Shaker song-writer visiting the 'happy region' containing the old settlements at Watervliet, New Lebanon and Hancock was inspired to write of how '[t]heir universal order appear'd a perfect charm both in the house and garden, the workshop and the farm' (in Patterson 1979: 145–6). Two writers summarise matters as follows:

> Shaker villages were characterised by neatness and extreme simplicity. Architecture was unadorned. 'Beadings, mouldings and cornices, which are *merely for fancy* may not be made by Believers', reminded the Millennial Laws of 1845. 'Odd or fanciful styles of architecture' were likewise prohibited.
>
> (Sprigg and Larkin 1988: 33)

Dolores Hayden discusses the Shaker settlement of Hancock, initially established in the 1790s, which she describes as having been marked by an extreme (even tedious) regularity of design that meant every building had its definite form and use, and that all the buildings together fitted into a highly regimented patterning:

> The 'order and use' or organisation and activities of the Hancock settlement were extremely legible because of the right-angle alignment of landscape boundaries and buildings, the hierarchical positioning of buildings according to functions, the distinctive shapes accorded to special functions, and the colour coding prescribed by the Millennial Laws. Barns and service buildings were dark colours, deep reds or tans; workshops and dwelling houses slightly lighter in colour,

36

yellows or creams; and the meeting house, white. Meticulous fencing, in iron, stone and wood, defined the boundaries of the Hancock domain and emphasised the pattern of land use.

(Hayden 1976: 77–81)

The Hancock community comprised six Shaker families, each of which occupied a collection of small wooden yellow or cream houses set at right angles to the main road through the settlement, and it appears that the outdoor public spaces – where much of the everyday interaction of an ordinary community would occur – were themselves designed to cut down on too much casual contact between individual Believers. These spaces were fenced square, and tight rules governed how people were supposed to move through them: there was a ban on 'cutting corners', for instance, and on loitering in such spaces when folk should be about their daily business (chatting at corners, sitting on fences, and the suchlike were out of the question). In stark contrast to the ramshackle spaces created by the Shaker Laners, what is remarkable about this micro-spatial organisation of the Shaker world is its thoroughgoing geometricism, its pervasive attention to right angles, straight lines, continuity of lines, proportion, subtle gradations in sizes: in short, to the 'purity of form and line' (Armstrong, in Sprigg 1986: 6).[17] Pictures and maps of settlements such as Hancock (see Figure 2.2) clearly capture this quality of geometric order, and in fact the cartographic representations sometimes played up the geometric regularity through the excessive use of rulers and set squares.[18] There is much that could be said about this straightened and gridded 'other rural', not the least being the possible dislocation between this geometricism and what might have been expected – a more fluid, sinuous and uncompartmentalised approach to space, perhaps – given the supposed prominence of 'feminine' values in the Shaker world-view.[19]

Furthermore, it seems that this aspect of Shaker spatial organisation extended to the still smaller spaces inside buildings, and hence into the control of highly personal geographies entailing bodily movements and interpersonal contacts, often in a manner producing a stark gendering of interior spaces and their utilisation. Every dwelling was rigidly divided into spaces for men and spaces for women, of course, and involved the so-called 'double circulation system' predicated on having separate staircases for the sexes and separate communal bedrooms on opposite sides of the corridors on the second and third floors.[20] At the same time definite rules were laid down about 'orientation and posture' of the body. Most notable of all – and as already hinted at – was the insistence on orthogonal forms (from settlement plans to building designs to the command that 'bread and meat are to be cut square') and orthogonal movements (from Believers not being allowed to take diagonal shortcuts to their not being able to pass serving dishes diagonally across the dinner table). Slouching

Figure 2.2 A Shaker settlement

was frowned upon because it spoiled the straight lines of the body, and when the Shaker went to bed he or she was supposed to 'retire to rest in the fear of God . . . and lie straight' (in Hayden 1976: 82). Hayden refers to what she terms the 'envelopes of space' that were imagined to surround each Believer, and then records certain 'distancing regulations' which demanded not only a separation of the sexes but also that a constant distance be maintained between Believers and non-Believers, such that when out in the world Shakers were discouraged from sitting close to non-Believers, from shaking their hands, and from permitting non-Believers to come between two Shakers, who together were reckoned to be properly enclosed by a 'double spatial envelope'. In this latter respect Shakers were advised that 'when you walk in the streets, you should keep so close together that there would not be room for even so much as a dog to run between you and your companion' (in Andrews and Andrews 1982: 42). It was only during events of worship, when whole Shaker communities came together in their meeting-houses to pray and to negotiate their individual and collective relationships with God, that these intense micro-spatial regulations were put aside. Indeed, many meetings became remarkably fluid, emotion-charged and even 'hysterical' affairs at which all kinds of motions, touches, sounds, wailings, chantings, dancings and ravings were likely to occur, and it is possible to speculate about the importance of these events in giving a 'space' for the release of the psycho-sexual desires that were surely being repressed much of the time in the lives of most Believers. And yet even here the Shaker elders sought to enforce a measure of regulation over the bodies and bodily spaces involved, and it was Father Joseph who converted the simple folk-dance forms favoured by the early Shakers – along with their whirlings, leapings and shakings – into 'spiritual exercises' which 'g[a]ve shape to ecstatic bodily operations' (Whitson 1983: 17) and entailed 'a simple, uniform dance that all Believers could practice as one, stepping forward and back in perfect unison' (Sprigg and Larkin 1988: 23). Even the most intimate of personal geographies were hence not free from the will to regulate, to create an order imagined as the gift of heaven, that was so ingrained in the thinking of the Shaker superiors.

BY WAY OF CONCLUSION: WHOSE 'OTHER RURALS'?

The specific inspiration for my chapter – and for its title – has been a brilliant short essay by Michel Foucault entitled 'Of other spaces', which was translated from the French and published in 1986, and which has since gained attention from several geographers (especially Soja 1989: 16–19). Foucault's chief aim in this essay is to flag the idea that the West's encounter with 'space' has a history, by which he means that the prevailing conceptual frameworks shaping people's experience and interpretation of space have changed quite dramatically over time, thus implying that 'the space which today appears to form the horizon of our concerns, our theory, our systems' (Foucault 1986: 22) is itself a human construction – an invention, albeit one with deep historical roots – quite different from earlier senses of what space entailed, comprised or signified. For Foucault, a medieval era energised by 'the space of emplacement' (a sense of spaces linked together in an overall hierarchy placing each in terms of its values and properties relative to all others) gave way to an early modern era awakening to 'the space of localisation' (a sense of spaces defined by their 'point' locations within broader patterns of movement specified by the laws of Galilean and Newtonian science), and then to a modern era predicated on 'the space of extension' (a sense of spaces inextricably bound into wider networks, systems and structures relating everything to everything else). What is relevant here is not so much the accuracy or otherwise of Foucault's history, however, as simply to stress the variability in human apprehensions of what all this spatial stuff really is: material and immaterial, existing and subsisting, around, near, by, through and even in 'us'. There are indeed 'other spaces', then, and Foucault forces his readers to begin thinking – or perhaps imagining is a better term – along quite new and different avenues (ones that question the tried and trusted, the expected, 'the Same') in the quest for fresh possibilities of spatial knowing (ones attending to calls from 'the Other'). The purpose of my chapter is less grand, but it does occur to me that there is a warrant for rural geographers adopting the spirit of Foucault's musings about 'other spaces' in the narrower task of contemplating the possibilities of 'other rurals'.

I would contend that this is a task which cannot be approached only through theoretical reflections, whether phenomenological, political-economic or whatever, since it is a project demanding attentive engagements with all manner of substantive 'other rurals' present in the total historical-geographical record of human dealings with the rural (from Dark Age forests to late-twentieth-century countryside theme parks, from English fens to the Australian outback, from familiar rurals here and now to unfamiliar rurals there and then). And I would argue that on various counts the Shakers in the rural United States managed to produce an 'other rural' which might strike many academic rural geographers today as rather strange, and that they do in consequence offer a stimulating case study in the Foucauldian project just outlined. Consider their muddling up of rural and urban visions; their treatment of rural locations in the context of seeking to place heaven on earth; their mobilisation of a rural–agricultural environment as a moralising resource; their all-pervading regulation of nature in an attempt to inscribe heavenly value into the debasement of brute earth; their minutely detailed organisation of space at scales ranging from the settlement plan to the dancing body; or their obsession with a geometric order which created a rural landscape 'unnaturally' traversed by straight lines and right angles (notwithstanding the supposed significance of a 'feminine' contribution to the running of the sect). In all of these ways (and probably more) the Shakers were surely 'rural others' actively producing an 'other rural', and in fact Foucault can also be a guide in this specific respect since – in the course of speculating about the reality of what he refers to as 'heterotopias', spaces that are 'something like counter-sites . . . in which the real sites, all the other sites that can be found within [a given] culture, are simultaneously represented, contested and inverted' (Foucault 1986: 24)[21] – special mention is made of Puritan settlements in New England and Jesuit colonies in South America as 'heterotopias' for the European societies that spawned them.[22] This implies that Foucault himself would have regarded the Shakers in the countryside as creating mini-'heterotopias' locked into a complex relation with the secular society of the eastern United States, comprising curious but appealing 'other spaces' which were in effect mirrors held up to the usual spaces and social lives of this society.

It is with Foucault's final thoughts about 'heterotopias' (see also Soja 1990) as surfaces in which 'the Same' sees itself reflected in 'the Other' that I will now conclude. To quote Foucault,

> The last trait of heterotopias is that they have a function in relation to all the space that remains. This function unfolds between two extreme poles. Either their role is to create a space of illusion that exposes every real space, all the sites inside of which human life is partitioned, as still more illusory (perhaps that is the role that was played by those famous brothels of which we are deprived). Or else,

on the contrary, their role is to create a space that is other, another real space, as perfect, as meticulous, as well arranged as ours is messy, ill-constructed and jumbled. This latter type would be the heterotopia, not of illusion, but of compensation.

(Foucault 1986: 27)

The Shaker settlements can evidently be understood as comprising the second type of 'heterotopia' described here, in which mainstream society sees its own obsessions with order racheted up to a higher and more intense level; and it might also be appropriate to conceive of the Shaker Lane settlement of the children's book as comprising the first type of 'heterotopia', in which mainstream society sees its attempts at creating order – its invention and policing of categories, boundaries and 'partitions' – mocked as illusory, deluded and futile.[23] The plausibility of such equations could be debated at length, but so too could the wisdom of adopting Foucault on 'other spaces' and 'heterotopias' as a guide in the exploration of 'other rurals'. My own feeling is that it will be highly instructive to do so, provided that the effort is laced with other theoretical perspectives on difference and marginality as proposed elsewhere in this volume, but this is not to suggest that Foucault's way of proceeding is beyond criticism. In this respect let me echo what several commentators have now identified as an element of ethnocentricism or Eurocentricism in Foucault's writings (e.g. Gregory 1994: 29, 190–3), and suggest that this problem does arise to an extent in his discussion of 'other spaces' because these spaces are consistently positioned as 'other' to a quite specific set of assumptions and behaviours typical of the powerful and the educated in the West. Despite Foucault's claims about how gazing on 'other spaces' (and more specifically on 'heterotopias') forces gazers to focus back upon themselves, to learn about themselves from the mirror images confronting them of their own worlds either hopelessly broken down or vigorously hyped up, he does not entertain the question of what happens if the gazers are not themselves conventional representatives of a middle-class, white, male, heterosexual, fit and healthy elite.[24] What 'other spaces' or even 'other rurals' would be perceived by either a Shaker elder or a resident of Shaker Lane, for instance, and how would such constructs then be reflected back into the thoughts and actions of these individuals? And could it be that what I personally identify as an 'other space' or an 'other rural' will be for many people different from me really quite unexceptional, and certainly not warranting any expressions of surprise?[25] And could it be that, even within the community of academics and even within that part of the community called geographers, what I take to be an 'other space' or an 'other rural' will not be so regarded by colleagues with different concerns and working from different traditions, institutions, backgrounds and parts of the world?[26] Doubtless in many cases my 'other rurals' would turn out to be ones

41

already integral to the imaginings of other people, whether inside the academy or without, and this means that the project of finding, exploring and representing 'other rurals' – vital as such a task surely is – will always need to be undertaken with the greatest of care.

ACKNOWLEDGEMENTS

Versions of this chapter have been given to a one-day conference of the Rural Economy and Society Study Group (on 'Problems of Marginalisation and the Representation of 'Others', Department of Planning and Landscape, University of Manchester, 25 May 1994), and also to seminars in both the Department of Geography and Topographic Science, University of Glasgow, and the Department of Geography, University of Aberdeen. Thanks are due to participants on these occasions for their thoughts and encouragement. More particularly, thanks are due to the editors, Paul Cloke and Jo Little, and also to Andy Cumbers, Alex Hughes, Catherine Nash, Hester Parr, Jo Sharp, Mark Shucksmith, Paul Routledge and Ian Thompson for their encouragment and suggestions.

NOTES

1 In particular, the Shaker Lane example reminds me of a paper that I heard some years ago (White 1986), which discussed how the 'humble distraught landscapes' of rural folk in the Willamette Valley – folk whose lands were characterised by a jumble of rusting machinery, chicken runs and vegetable gardens – were coming under threat from a ruralising urban population with very different ideas about how the countryside should be ordered and managed. Other geographical works on such 'outsiders' include Sibley (1981, 1992, 1994), and also Kearns (1977, 1978) and Sinclair (1993).

2 In their thoughtful responses to my paper Murdoch and Pratt (1993, 1994) are more certain that rural studies is in a need of a thoroughgoing transformation, one inspired by the claims of postmodernism and other (potentially) 'critical' social theories. Jones usefully contrasts '[the] more radical paradigm shifts advocated by Murdoch and Pratt' with 'the more intimate small-scale sensitising of existing approaches as advocated by Philo' (Jones 1995: 37). I realise that some may see a contradiction between my own reluctance to dismiss wholesale the existing corpus of rural studies and my urging here of rural geographers to abandon the prison-house of 'the Same' in seeking out the new conceptual–substantive possibilities of 'the Other', but I actually have no difficulty with the notion of orthodox and alternative approaches co-existing within rural research (why does it have to be a case of 'either/or'?).

3 Olsson has long been preoccupied with precisely this issue as well: that of how the conventional processes of conceptualisation deeply ingrained in Western thought continually work to capture, to locate, to put boundaries around the things of the world as they enter into the immateriality of human minds, and as they are then represented through acts of speaking and (more particularly) writing. He hence argues about the 'invisible geographies' or 'cartographies' of thought, the configurations of lines and shapes which are seemingly indispens-

able to Western reason, and wonders about the possibilities for quite other geographies of thought which operate to – or, better, serve to install – quite different logics in which such devices as 'either/or' are replaced by senses of 'and, and, and . . .' (see Olsson 1980, 1991). Olsson's high-modernist experiments in this respect are arguably energised by similar frustrations and ambitions, and perhaps too by much the same understanding of 'geography', as the post-modernist–deconstructionist efforts of someone like Doel (e.g. 1993, 1994a, 1994b). In my case study of the Shakers, it could of course be claimed that – thanks to my using a conventional academic layout, neatly subsectioned and logically argued through – I have not really challenged conventional intellectual ordering principles, and have thereby allowed my account to remain firmly locked within the Western academy's 'apparatuses of capture' (Paul Routledge, in seminar discussion).

4 An intriguing argument for retaining a focus on 'rural sames', and for recognising that even here the coherence or predictability of people's constructions of the rural may not be as great as might initially be supposed, is contained in Halfacree (1995). He situates his argument in the context of debates about the necessity of looking more closely at 'rural others' (Philo 1992, 1993; Murdoch and Pratt 1993, 1994). A rather different but highly instructive attempt to excavate the senses of rurality possessed by 'rural sames' is written through in Matless (1993).

5 A warning might be made here about the totalising pretensions of a concept such as 'the Other', which itself risks becoming an all-embracing category identical in principle to the grand-theoretical devices it is supposed to undermine: one that gathers to itself myriad dimensions of substantive otherness, stitching together the diversity of existing human 'others' in the process, and perhaps thereby replicating how 'the Same' ignores, obliterates and fails to attend to the many differences of the human world (see also Duncan and Sharp 1993). I should add that there are a number of different concepts of 'the Other' in circulation, relating to Mead's social psychology, Lacanian psychoanalysis and presumably other intellectual traditions as well.

6 I must thank Catherine Nash for allowing me to consult materials that she has collected on the Shakers in the context of her own interest in 'utopian' thought and practice (e.g. Nash 1994). From her and through my own library work, I have now read quite widely in the small corpus of book-length Shaker scholarship, and in what follows I draw upon Andrews and Andrews (1982), Brewer (1986), Desroche (1971), Faber (1974), Horgan (1982), McGuire (1989), Melcher (1941), Neal (1947), Patterson (1979), Rubin (1984), Sasson (1983), Sprigg (1986), Sprigg and Larkin (1988), Stein (1992), Whitson (1983) and Whitworth (1975).

7 The theme of 'settlers' encountering the North American frontier, with special reference to the cultural landscapes produced in the process, has long been popular in geographical inquiry (a classic paper in this respect is Trewartha 1946). More particularly, research has been conducted on the specific patterns of settlement on the frontier (and indeed elsewhere in rural North America) associated with distinctive ethnic–religious groupings such as the Amish, Jews, Mennonites, Mormons and early Puritans (e.g. Crowley 1978; Francaviglia 1970, 1971; Lehr and Katz 1995; Warkentin 1959). In addition, various suggestions have been made about the connections between the geographies of religious groupings, the areal patterning of social–cultural phenomena, and the specific impact of religious practices on local landscapes (e.g. Ley 1994; Sopher 1967; Zelinsky 1961). I have found no mention of the Shakers in any of this geographical literature.

8 Feminist geographers have demonstrated that within Western societies a division between 'men's space' an 'women's space' often overlaps with that between 'public space' and 'private space', an overlap bound up in part with the conventional stress on the nuclear family unit, and they have then speculated about the potential of adopting more collective or communal living arrangements to achieve a blurring of private and public which would at the same time be a fusing of male ('productive') and female ('reproductive') spaces (see especially McDowell 1983).

9 One writer describes this site of the first Shaker settlement as 'mostly swampy wilderness', and notes that – notwithstanding the presence of Albany, a sizeable centre only seven miles to the south-east – 'it had such a number of low and swampy spots that other people had spurned it in favour of higher and healthier locations' (Faber 1974: 30–3).

10 Desroche casts the Shakers as 'one of the first chapters in the prehistory of modern socialism' (Desroche 1971: 3 and especially ch. 7), noting linkages with certain streams of 'radicalism' (those of Owen and Tolstoy, for instance, if not Marx) and also describing the working-class origins of the sect as bound up with hints of a 'republican' political consciousness. Other writers are suspicious of such claims, attacking accounts which present Shakerism 'as a half-way house to Marxian materialism' (Patterson 1979: xv) and preferring instead to regard it as 'a form of people's capitalism ... in which every adherent had a stake, economic or religious' (Andrews and Andrews 1982: 45).

11 In this respect, it has been noted that the New England Shakers 'were spared the preliminary task of clearing all their land, since many of the early converts in Massachussetts, New Hampshire and Maine brought already cultivated lands and houses already built into joint ownership' (Melcher 1941: 120). Horgan's study of the Harvard and Shirley settlements in Massachusetts reveals that 'Harvard's rural countenance was placid', and indicates that even before the Shakers arrived, 'farms and orchards, gristmills and sawmills, cottage industries and small family holdings were sparsely distributed over Harvard's hills and valleys' (Horgan 1982: 20). What Horgan's study also reveals, though, is that existing 'townspeople' were very hostile to the Shakers and tried to chase them out of the district on several occasions.

12 The task facing the Shakers on the frontier was very different from that of their Eastern counterparts, since they had to endure 'all of the struggles of pioneers in a new country where nature is the first foe to be subdued' (Melcher 1941: 120). The life of Shakers on the western frontier is a common theme in the Shaker literature, and one work specifically charts the frontier experience of the South Union settlement (initially the Gasper 'mission station') in Kentucky, first established 1807–1813 (see Neal 1947). The frontier provided a grouping such as the Shakers with 'the opportunity to acquire land quickly, with secure titles and democratic freedom to arrange their life as they wished.... Whether [they] "acquired" space by conversions or legal space by purchase ... this secure legal status must have been important to the survival of the sect and allowed them to protect their "otherness" initially' (Ian Thompson, pers. comm.). Furthermore, there may have been a connection between the geometric layout of Shaker settlements and the rectangular lots (the 'legal space') in which land was sold off, although it is also the case that this geometric imprint appeared even when they took up lands already boasting a history of settlement.

13 In this regard the Shakers often became quite prosperous, despite not really aiming to make profits, which means that they were a 'rural other' for whom cultural marginality did not necessarily match up with economic marginality.

14 One compelling song in this respect was adapted by the Believers of Enfield, New Hampshire, from an Indian chief called Contoocook (indicating an intriguing line of 'hybridity' that could emerge on occasion):

> Me love me hills and mountains,
> Me love me pleasant groves,
> Me love to ramble around as me feel as me choose,
> Me love me pleasant waters,
> To bathe in their pure streams,
> And fish me on the banks,
> Or in me canoe to skim.

(in Patterson 1979: 353)

15 It is suggested that '[a]ll through their existence they have remained faithful to the land – an agricultural people living by necessity and preference in country districts' (Melcher 1941: 120).

16 One writer states that 'the Shakers regarded themselves as God's stewards and pioneers, and their labour as an act of worship and of reclamation. To fertilise the soil of the communities was to improve the land which, in an especially intimate way, belonged to God, and so was in some part a furtherance of God's plan' (Whitworth 1975: 34).

17 This concern for the 'purity of line and form' is most commonly noticed in connection with Shaker furniture, and – writing more generally about Shaker design – Sprigg concludes that '[w]hat really distinguishes Shaker design is something that transcends utility, simplicity and perfection – a subtle beauty that relies almost entirely on proportion. There is harmony in the parts of a Shaker object' (Sprigg 1986: 11). For Sprigg, there is evidently nothing contradictory about geometric precision and the creation of 'a simple beauty'.

18 It is remarked that '[w]hile art for its own sake was not acceptable to the Shakers . . . Believers encouraged the making of maps of their own villages. Such views and plans were useful for community planning, and perhaps also served a spiritual purpose as records of these heavens on earth' (Sprigg and Larkin 1988: 50).

19 This is a controversial claim, of course, since it teeters on the brink of an essentialism implying that women and men have fundamentally different senses of space. This being said, feminist geographers such as Rose are wondering about new senses of space – new spatial metaphors to deploy – which are intimately bound up with the 'tactile spatialities' of women's bodies, perhaps scrambling notions of inside and outside, inclusion and exclusion, surface and depth (e.g. Rose 1991, 1995: see also Irigaray 1991). The parallel argument is that the conceptions of space which prevail within much of Western society are the product of a patriarchal hegemony, one that denies the presence of bodies, passions and emotions, and one that in so doing creates 'masculinist' abstract spatialities predicated on lines, forms and projections – the elements of conventional geometry – which are in effect 'unnatural' (artificial, imposed, controlling), against, above or outside of nature.

20 An intriguing contrast exists here with Mormon dwellings, since – whereas the Shaker house was designed to prevent the sexes mixing and entering into sexual relations – the Mormon 'I'-shaped house was designed to facilitate just such activity: 'it is sometimes called a "polygamy house". . . . The early Mormons may have found this house useful for polygamy, since it was symmetrical and sometimes divided into two equal halves, each having a front door with no

interior communication between the two halves of the house' (Francaviglia 1971: 65).

21 Foucault distinguishes between 'utopias' and 'heterotopias' chiefly on the grounds that the former do not exist – they are 'sites with no real place', 'fundamentally unreal spaces' (Foucault 1986: 24) – whereas the latter are actually existing in the world, serving as half-way houses towards utopia or indeed dystopia, real spaces where attempts have been made (whether deliberately or not) to produce what might end up being regarded as either heaven or hell on earth.

22 Talking of the villages constructed by the Puritans and the Jesuits, Foucault writes as follows: '[t]he village was laid out according to a rigorous plan around a rectangular place at the foot of which was the church; on one side, there was the school; on the other, the cemetery; and then, in front of the church, an avenue set out that another crossed at right angles; each family had its little cabin along these two axes, and thus the sign of Christ was exactly reproduced. Christianity marked the space and geography of the American world with its fundamental sign' (Foucault 1986: 27). The ubiquity of the geometric precision to such religious settlements is here emphasised, and a grand hypothesis is proposed about why this geometricism – this obsession with straight lines and right angles – may have emerged.

23 Given that Shaker Lane is only a fictional case, though, it cannot really be described as a Foucauldian 'heterotopia'. It is intriguing to wonder why Provensen and Provensen (1991) chose to call their settlement 'Shaker Lane', and to wonder whether they intended any association with a grouping which – despite its tendency towards being hyper-ordered – would almost certainly be regarded by mainstream society today as hardly less 'disordered' than the Shaker Laners (a point suggested to me by Paul Routledge).

24 Foucault is clearly aware that he is considering the 'other spaces' of the West (and more particularly of Western Europe), but the possibility of differences within the 'we' that is here looking out on the world – scanning this world for its 'other spaces' and 'heterotopias' – is not acknowledged. There is no deeper problematising of the 'us' which is positioned as the identifier of 'other spaces', no sense that 'we' might not be all the same, no recognition that 'we' might not be anything like in agreement about what is to be regarded as 'other', different, strange. This is a key claim for my chapter, and one that does rebound on me and the implication that 'us' – as academic rural geographers – will be able to speak of 'other rurals'. In response, though, I would suggest that to speak of 'others' (to contemplate, with Foucault, the possibilities of 'other spaces' and even 'other rurals') is not necessarily to fall into an ethnocentricism, provided that the academics involved in such a project remain constantly alert to the many possible relativities involved in who regards whom as 'other' and from which places. Duncan and Sharp (1993) offer a broader perspective on such matters in seeking to salvage a notion of 'the Other', one alert to the meeting of different discourses and representations emanating from different 'subject positions' around the world, and they propose that such an approach can prevent academics becoming complicit with an 'Othering' process that fixes an undifferentiated 'us here' in opposition to an equally undifferentiated (and maybe even inferiorised) 'them there'.

25 A thoughtful auto-ethnographic account of one rural geographer's senses of what constitutes the rural can be found in Cloke (1994), in the course of which the author critically reflects upon both personal experiences and the intellectual milieux shaping what he sees as otherness and sameness in the countryside (and

which in effect influence what he regards as an 'other rural' or even a 'same rural'). Cloke echoes points made above when underlining the problem of 'the researcher [who] is too busy looking for sameness', and who thereby fails to be open to 'the recognition of other geographies of the rural' (Cloke 1994: 162), but what he further claims – surely with some justification – is that dangers attach to an unthinking pursuit of 'neglected rural geographies' which fashionably seeks out 'particular facets of rural difference (gender, ethnicity, alternativeness, etc.) while other rural groups will be left out' (Cloke 1994: 183; see also Halfacree 1995). The personal and intellectual motivations leading rural geographers to identify and to speak of 'other rurals', and to claim any kinds of academic or 'political' kudos in the process, should always be considered (not least by the individuals themselves).

26 This is something that became apparent to me on moving to take up a new position (in Glasgow), and finding that among my new colleagues were geographers whose senses of the rural have been shaped by experiences of French, South American, African and Middle Eastern rural–agricultural environments (e.g. Briggs 1991; Thompson 1995). My personal 'other rurals' might therefore turn out to be rurals with which they are quite familiar, and which they might not regard as particularly 'other'.

REFERENCES

Andrews, E. D. and Andrews, F. (1982) *Work and Worship among the Shakers: Their Craftsmanship and Economic Order*, New York: Dover Publications.

Bell, D. and Valentine, G. (1995) 'Queer country: rural lesbian and gay lives', *Journal of Rural Studies* 11: 113–22.

Brandth, B. (1995) 'Rural masculinity in transition: gender images in tractor advertisements', *Journal of Rural Studies* 11: 123–33.

Brewer, P. J. (1986) *Shaker Communities, Shaker Lives*, London: University Press of New England.

Briggs, J. (1991) 'The peri-urban zone of Dar es Salaam, Tanzania: recent trends and changes in agricultural land use', *Transactions of the Institute of British Geographers* NS 16: 319–31.

Cloke, P. (1994) '(En)culturing political geography: a life in the day of a "rural geographer"', in Cloke, P., Doel, M., Matless, D., Phillips, M. and Thrift, N. *Writing the Rural: Five Cultural Geographies*, London: Paul Chapman.

Cloke, P. and Thrift, N. (1990) 'Class and change in rural Britain', in Marsden, T., Lowe, P. and Whatmore, S. (eds) *Rural Restructuring: Global Processes and Their Responses*, London: David Fulton.

Cross, W. R. (1950) *The Burned-Over District: The Social and Intellectual History of Enthusiastic Religion in Western New York, 1800–1850*, Ithaca: Cornell University Press.

Crowley, W. K. (1978) 'Old Order Amish settlement: diffusion and growth', *Annals of the Association of American Geographers* 68: 249–64.

Davey, J. A. and Kearns, R. A. (1994) 'Special needs versus the "level-playing field": recent developments in housing policy for indigenous people in New Zealand', *Journal of Rural Studies* 10: 73–82.

Desroche, H. (1971, trans.) *The American Shakers: From Neo-Christianity to Presocialism*, Amherst: University of Massachusetts Press.

Doel, M. A. (1993) 'Proverbs for paranoids: writing geography on hollowed ground', *Transactions of the Institute of British Geographers* NS 18: 377–94.

—— (1994a) 'Deconstruction on the move: from libidinal economy to liminal materialism', *Environment and Planning A* 26: 1041–59.

—— (1994b) 'Something resists: reading-deconstruction as ontological infestation (departures from the texts of Jacques Derrida)', in Cloke, P., Doel, M., Matless, D., Phillips, M. and Thrift, N. *Writing the Rural: Five Cultural Geographies*, London: Paul Chapman.

Duncan, N. and Sharp, J. P. (1993) 'Confronting representation(s)', *Environment and Planning D: Society and Space* 11: 473–86.

Faber, D. (1974) *The Perfect Life: The Shakers in America*, New York: Farrar, Straus and Giroux.

Foucault, M. (1967, trans.) *Madness and Civilization: A History of Insanity in the Age of Reason*, London: Tavistock.

—— (1970, trans.) *The Order of Things: An Archaeology of the Human Sciences*, London: Tavistock.

—— (1972, trans.) *The Archaeology of Knowledge*, London: Tavistock.

—— (1986, trans.) 'Of other spaces', *Diacritics* 1, Spring: 22–7.

Francaviglia, R. V. (1970) 'The Mormon landscape: definition of an image in the American West', *Proceedings of the Association of American Geographers* 2: 59–61.

—— (1971) 'Mormon central-hall houses in the American West', *Annals of the Association of American Geographers* 61: 65–71.

Gregory, D. (1994) *Geographical Imaginations*. Oxford: Blackwell.

Halfacree, K. H. (1995) 'Talking about rurality: social representations of the rural as expressed by residents of six English parishes', *Journal of Rural Studies* 11: 1–20.

Harrison, J. F. C. (1979) *The Second Coming: Popular Millennarianism, 1780–1850*, London: Routledge and Kegan Paul.

Hayden, D. (1976) *Seven American Utopias: The Architecture of Communitarian Socialism*, Cambridge, Mass.: MIT Press.

Horgan, E. R. (1982) *The Shaker Holy Land: A Community Portrait*, Harvard, Mass.: Harvard Common Press.

Irigaray, L. (1991, trans.) 'Sexual difference', in Whitford, M. (ed.) *The Irigaray Reader*, Oxford: Basil Blackwell.

Jones, O. (1995) 'Lay discourses of the rural: developments and implications for rural studies', *Journal of Rural Studies* 11: 35–49.

Kearns, K. C. (1977) 'Irish tinkers: an itinerant population in transition', *Annals of the Association of American Geographers* 67: 538–48.

—— (1978) 'Irish tinkers are unwelcome', *Geographical Magazine* 50: 329–34.

Lehr, J. C. and Katz, Y. (1995) 'Crown, corporation and church: the role of institutions in the stability of pioneer settlements in the Canadian West, 1870–1914', *Journal of Historical Geography* 21: 413–29.

Ley, D. (1994) 'Religion, geography of', entry in Johnston, R. J., Gregory, D. and Smith, D. M. (eds) *The Dictionary of Human Geography, 3rd edition*, Oxford: Blackwell.

McDowell, L. (1983) 'Towards an understanding of the gender division of urban space' *Environment and Planning D: Society and Space* 1: 59–72.

McGuire, J. (1989) *Basketry: The Shaker Tradition – History, Techniques, Projects*, New York: Sterling.

Mackenzie, F. (1994) ' "Is where I sit where I stand", the Ontario Farm Women's Network, politics and difference', *Journal of Rural Studies*, 10, 110–16.

Matless, D. (1993) 'Doing the English village, 1945–1990: an essay in imaginative geography', in Cloke, P., Doel, M., Matless, D., Phillips, M. and Thrift, N. *Writing the Rural: Five Cultural Geographies*, London: Paul Chapman.

Megill, A. (1985) *Prophets of Extremity: Nietzsche, Heidegger, Foucault, Derrida*, London: University of California Press.

Melcher, M. F. (1941) *The Shaker Adventure*, Princeton: Princeton University Press.

Murdoch, J. and Pratt, A. C. (1993) 'Rural studies: modernism, postmodernism and the post-rural', *Journal of Rural Studies* 9: 411–27.

—— (1994) 'Rural studies and power and the power of rural studies: a reply to Philo', *Journal of Rural Studies* 10: 83–7.

Nash, C. (1994) 'Utopia, gender and regeneration: Irish utopian geographies and restoration of the masculine, 1900–1920', paper given at the Annual Meeting of the Institute of British Geographers, 4–7 January 1994.

Neal, J. (1947) *By Their Fruits: The Story of Shakerism in South Union, Kentucky*, Chapel Hill: University of North Carolina Press.

Olsson, G. (1980) *Birds in Egg/Eggs in Bird*, London: Pion.

—— (1991) *Lines of Power/Limits of Language*, Minneapolis, University of Minnesota Press.

Patterson, D. W. (1979) *The Shaker Spiritual*, Princeton: Princeton University Press.

Philo, C. (1986) ' "The Same and the Other": on geographies, madness and outsiders', Loughborough University of Technology, Department of Geography Occasional Paper No. 11.

—— (1992) 'Neglected rural geographies: a review', *Journal of Rural Studies* 8: 193–207.

—— (1993) 'Postmodern rural geography? a reply to Murdoch and Pratt', *Journal of Rural Studies* 9: 429–36.

Provensen, A. and Provensen, M. (1991) *Shaker Lane*, London: Walker Books.

Rose, G. (1991) 'On being ambivalent: women and feminisms in geography', in Philo, C. (comp.) *New Words, New Worlds: Reconceptualising Social and Cultural Geography*, St David's University College, Lampeter: Social and Cultural Geography Study Group.

—— (1995) 'Tradition and paternity: same difference', forthcoming in *Transactions of the Institute of British Geographers* 20: 414–16

Rubin, C. E. (1984) *Shaker Herbs: An Essay by Cynthia Elyce Rubin with Nineteenth-Century Shaker Herb Labels*, Northampton, Mass.: Catawba Press.

Sasson, D. (1983) *The Shaker Spiritual Narrative*, Knoxville: University of Tennessee Press.

Sibley, D. (1981) *Outsiders in Urban Societies*, Oxford: Basil Blackwell.

—— (1988) 'Purification of space', *Environment and Planning D: Society and Space* 6: 409–21.

—— (1992) 'Outsiders in society and space', in Anderson, K. and Gale, F. (eds) *Inventing Places: Studies in Cultural Geography*, Melbourne: Longman Cheshire.

—— (1994) 'The sin of transgression', *Area* 26: 300–3.

—— (1995) *Geographies of Exclusion: Society and Difference in the West*, London: Routledge.

Sinclair, P. (1993) 'Casting out the outcasts', *Geographical Magazine* 65: 14–18.

Sopher, D. (1967) *Geography of Religions*, Englewood Cliffs: Prentice-Hall.

Soja, E. W. (1989) *Postmodern Geographies: The Reassertion of Space in Critical Social Theory*, London: Verso.

—— (1990) 'Heterotopologies: a remembrance of other spaces in the Citadel–LA', *Strategies* 3: 6–39.

Sprigg, J. (1986) *Shaker Design*, New York: Whitney Museum of American Art, with W. W. Norton & Co. (with a foreword by Armstrong, T.)

Sprigg, J. and Larkin, D. (1988) *Shaker: Life, Work and Art*, London: Cassell.

Stein, S. J. (1992) *The Shaker Experience in North America: A History of the United Society of Believers*, New Haven: Yale University Press.

Thompson, I. B. (1995) 'A "generational" approach to reconstructing rural change using oral evidence: a case study from Creuse, France, 1870–1972', in Pitte, J.-R. (ed.) *Géographie Historique et Culturelle de l'Europe: Hommage au Professeur Xavier de Planhol*, Paris: Presses de l'Université de Paris-Sorbonne.

Trewartha, G. (1946) 'Types of rural settlement in colonial America', *Geographical Review* 36: 568–96.

Warkentin, J. (1959) 'Mennonite agricultural settlements of southern Manitoba', *Geographical Review* 49: 342–68.

White, W. R. (1986) 'Folk places, popular culture and the law in Willamette counties', paper given at the Annual Meeting of the Association of American Geographers, 3–7 May 1986.

Whitson, R. E. (ed.) (1983) *The Shakers: Two Centuries of Spiritual Reflection*, London: SPCK (with introductory essay by Whitson)

Whitworth, J. M. (1975) *God's Blueprints: A Sociological Study of Three Utopian Sects*, London: Routledge and Kegan Paul.

Zelinsky, W. (1961) 'An approach to the religious geography of the United States: patterns of church membership in 1952', *Annals of the Association of American Geographers* 51: 139–93.

3

FROM THE POWER OF TOPOGRAPHY TO THE TOPOGRAPHY OF POWER

A discourse on strange ruralities

Jonathan Murdoch and Andy C. Pratt

INTRODUCTION

The terms 'rural' and 'countryside' tend to evoke images of harmony and consensus. In Britain such images derive much of their power from the proximity of the countryside ideal to British national identity, and it has always seemed to enshrine those timeless qualities that make this 'sceptred isle' forever 'England'. Rural land is considered a priceless part of the nation's heritage. It has traditionally been a 'cosy corner' in which an 'Anglocentric' culture, one opposed to the multiculturalism increasingly evident in many cities, could nestle down safe from harm (Lowe *et al.* 1995). A not dissimilar view of the countryside has often been (unwittingly) rehearsed in academic writing on rural Britain. Thus academic texts have frequently portrayed the rural as a homogeneous social space, one which seems in many ways to exist in some timeless zone where old-fashioned virtues and their associated forms of life still linger. And while in the past thirty years or so a more critical approach to the countryside has been evident, as recent debates, to be discussed below, have shown, even these accounts have sometimes thrown up new versions of the same old rural myths.

At the present juncture, and as the present volume indicates, a much more critical stance is being adopted towards such dominant images of rurality. Much academic writing currently emphasises how our received views of the rural landscape and county life disguise conflictual, competitive and exploitative sets of social relations. Thus the countryside that lies behind the images has become subject to a great deal of attention and, unsurprisingly perhaps, does not seem to quite match up to rural ideals (e.g. Thrift 1989). In this chapter we will briefly assess some of this work with the intention of sounding a note of caution. Before hurrying to the conclusion that we have now discovered some essential truth about the rural, we believe it is apposite to take note of Chambers' (1994)

51

comment on many current theoretical concerns: that what we might be witnessing is the same old academic 'will to power' extending its reach over new research subjects. This warning relates in part to the charge that is frequently levelled at academic writing, namely, that it simplistically seeks to sweep up the world into 'tidy' or, perhaps even more damning, 'abstract' concepts and categories. While such critiques are often over-drawn, we will argue below that this worry ought to be taken seriously and it should require that we look reflexively upon our own ways of doing things and should provoke us into continually monitoring the ways in which we conduct social science.

The aim of this chapter is to argue that a reflexive awareness should permeate current theoretical and methodological approaches in rural studies. It is our contention that such an awareness is increasingly neces-sary. Its necessity does not derive only from issues intrinsic to social science. The UK government has recently published its White Paper on the countryside; this represents the first wide-ranging survey of countryside and rural policies in the UK for fifty years (Department of the Environ-ment 1995). A major element in prompting this reappraisal has been the changing role and definition of the countryside within British society, as expressed in the competing and often conflicting demands made on rural resources. The White Paper, which is effectively an attempt to bring some co-ordination to the multitude of policies bearing upon rural Britain, reflects, in part, a fragmentation of 'the' rural experience. But its policies also represent a retreat from the governance of the rural as a coherent and unified space; the White Paper now proclaims that the value of the country-side lies in its diversity and local character and that its governance might be best achieved at the lowest level. Thus rural communities are asked to become more and more responsible for their own affairs, picking up many of the services and activities which were once the provenance of a (national) welfare state. We should be wary, then, of allying our new concerns with difference and diversity to particular policies at a time when government is seeking to retreat from many of its traditional (welfarist) responsibilities in rural areas. Such concerns should, therefore, be presented with an atten-tion to the sets of power relations that both surround and constitute academic discourses.

Attending to the configuration of power surrounding academic discourse has many implications for writing and researching. One aspect, which we wish to emphasise here, is the role of reflexivity in making arguments and proposals; that is, how can we critically interrogate, in an ongoing fashion, our ways of seeing and organising the rural world? What motivates our concern here is the belief that social scientific understandings of 'rurality' do matter; they matter because the way in which we 'see' the countryside affects perception (some aspects become 'visible', i.e. acknowledged as a problem and potentially amenable to modification, and others invisible, or overlooked), evaluation (the relative valuing or moral worth attributed to what is 'visible'), and policy proposals (the policies and the politics used to achieve particular ends).

In this chapter we also hope to clarify the contours of the 'new' ruralities that are emerging from recent work and highlight the extent to which these offer a radical challenge to the more traditional modes of analysis which are usually brought to bear upon the 'rural', a challenge which is captured in our title, drawn from a quotation by Gupta and Ferguson (1992: 8–9). In a critique of social anthropology they note 'the power of topography to conceal successfully the topography of power'. Gupta and Ferguson are referring to the way that traditional social anthropology has reified 'the village' as a bounded, sealed space and has ignored the flows of social relations across boundaries. We will have more to say about boundaries later in this chapter, but for now this quotation stands as a neat way of summarising our concerns about rural studies (by which we mean those academic disciplines concerned with the rural: geography, sociology, social anthropology, agricultural economics, etc.); namely, that the particular constructions of the rural that rural studies deploy – the topographies of the rural – have obscured or concealed that which sustains them – the topographies of power.

FROM POSTMODERN RURAL STUDIES TO 'POST-RURAL' STUDIES: A REVIEW

There has been considerable discussion recently within the pages of the *Journal of Rural Studies* on the implications of 'postmodernism' for rural studies (see Halfacree 1993; Murdoch and Pratt 1993, 1994; Philo 1993; Pratt 1996a, 1996b). This has been supplemented by a recent book (Cloke *et al.* 1994) which also draws upon a postmodernist sensibility to articulate a series of 'positions' on the rural. The flavour of these contributions, and the issues they raise, can be explored through an exchange between Chris Philo and ourselves (Philo 1992, 1993; Murdoch and Pratt 1993, 1994) in the aforementioned journal.

The article which more than any other prompted recent discussions about the scope and status of rural studies is Chris Philo's (1992) extended review of Colin Ward's book *Child in the Country*. In this piece Philo identifies a 'blind spot' in rural studies: that of an unwarranted focus upon, and concern with, the interests and activities of powerful groups within the countryside (e.g. middle-class incomers, farmers, planners). He claims that this oversight has resulted in the active exclusion of many other actors and social groups from the academic discourse. Philo seeks to formalise this critique using a term popularised in postmodern analyses: the Other (Philo has outlined his version of postmodern geography in Cloke *et al.* (1992)). 'Others' are those regarded as in some way illegitimate members of society as a result of a variety of social characteristics such as being gay, a single parent, a traveller, a black person and so on. Philo thus calls for rural geography to be open to diversity and urges geographers to explore a range of different 'voices' or experiences of the rural. A new rural studies can be fashioned which both recovers and includes a range of 'other' voices, voices which have previously been neglected.

In many ways Philo's concerns could be seen as congruent with quite long-standing research interests on the part of those traditionally working within the rural studies genre. For instance, a persistent concern within rural geography and sociology has been that of 'insider–outsider' relations, developed most notably within work on migration and counterurbanisation (see reviews by Halfacree 1995; Murdoch and Marsden 1994). Writers such as Pahl (1970) and Ambrose (1975), for instance, have sought to identify the divisions that have emerged within rural communities as a result of migratory (or counterurbanisation) trends. Although they used the concept of class (see Murdoch 1995 for an overview) as their main analytical tool, and thus concentrated upon class relations, they were nevertheless interested in how social groups in rural areas are changing as a result of dynamic processes of transformation. They were thereby continuing a tradition that can be traced from Wirth's (1938) classic study which questioned the *inter*-class difference between urban and rural spaces and suggested that *intra*-class differences may be just as significant. Ambrose and Pahl identified discrete types of 'incomers' in villages, some of whom have a pronounced impact on rural social formations, resulting in increased social tension. One important consequence of such analyses was that they effectively destroyed the idea that rural society is some kind of organic whole (it is riven by class relations at least) and sparked off a whole series of studies into the social processes that effectively transgress urban–rural distinctions (the work of Howard Newby in the late 1970s (Newby 1980; see also Newby *et al.* 1978) followed this line of analysis, as did much of the work on the political economy of agriculture during the 1980s (see Marsden *et al.* (1990) for a summary). Philo's concerns for rural Others could be simply taken as a rather belated acknowledgement that there is a greater diversity in the social relations which constitute the rural than has previously been recognised. However, Philo is concerned to explore a far more fractured set of identities than was evident in the earlier body of work and he urges researchers to be sensitive to the diversity of interests represented in the countryside.

We were stimulated to reply to Philo's article (Murdoch and Pratt 1993) for several reasons. First, we believe that the issues he addressed are significant and should not be ignored. Second, we agree with his identification of a partial rural geography, skewed in its research priorities. Third, we feel that his call for an epistemological engagement with postmodernism (spelled out more clearly in Philo 1993) is timely. However, although we agree with the general thrust of Philo's critique, we are reluctant to go along wholeheartedly with his prescription for rural studies. We are particularly worried by his attempt to resolve the problems of rural geography by simply 'adding in' a concern with 'new voices' or identities, particularly if the conceptual and methodological tools of analysis are set to remain the same. Our concern is that this can easily be seen as an indication that rural studies can continue in time-honoured fashion. Yet at present it is unclear how difference, diversity and fragmentation can be understood, given our current frames of reference, and there is a real danger that these new concerns will simply be grafted on at the margins. In short, our worry is that hidden and neglected 'Others' may remain peripheral to a subdiscipline still oriented towards the modernisation or rational organisation of the rural sphere. In our view the only way to counter this threat effectively is by reconstituting the 'core' of the subdiscipline: that is, by turning to a consideration of the organising frameworks that make up rural studies/geography/sociology. We may then begin to understand how these 'Others' came to be 'Othered'. Thus, what is required is more than *bricollage*; it is a reflexive reconstruction of the academic terrain.

In our responses to Philo we proposed that there is a need to understand how we come to know the 'rural'. One answer to this question can be found, we argued, within an analysis of the theorisations and methodological tools deployed by researchers of the rural domain. We offered as evidence a short (and partial) history of rural studies. We were inspired in this approach by Foucault's earlier works, such as the *Archaeology of Knowledge* and *The Order of Things*, where the aim is to 'create a history of the different modes by which... human beings are made subjects' (Foucault 1982: 208). Rabinow (1986: 7–8), in a commentary on Foucault, summarises the impetus as an investigation of the objectification of subjects through 'dividing practices' or 'scientific classification'. In short, we identified structures of knowledge about the rural that have created particular effects. One of these effects is what comes to be regarded as within the legitimate scope of rural studies; analytical frameworks determine what is seen and what is not.

We offered the idea that the modernist frameworks that have traditionally been used to make sense of rural society have consistently 'purified' (Sibley 1988) rural space and have reproduced particular socio-spatial dualisms. There has been a tendency to create divisions, categories or classifications which in turn 'perform' (Callon and Law 1995) different (but closely related) versions of the rural (established against, and usually inferior to, the urban). As a result, particular practices, spaces and persons are constituted as (il)legitimate on the basis of these (usually purified) categories (evidence on this process in another, non-academic domain is provided by Halfacree (1996)). A similar observation is made by Marc Mormont (1990: 22) when he says of the rural:

The category that evolved was not only empirical or descriptive . . .
it also carried what I shall call a representation or set of meanings,
in that it connoted a more or less explicit discourse ascribing a certain
set of characteristics or attributes to those to whom it was applied.
As any social category in ordinary use likewise implicitly ascribes
properties to groups, the set of meanings underpinning it is necessarily linked to a representation of society overall.

Moreover, he goes on to say, 'such a category alludes not only to objective
conditions, but also to social legitimation' where a particular set of meanings will 'confer a greater or lesser degree of validity on each social group'.
On these grounds, we should be extremely wary of attempts to definitively
define the rural. It is perhaps preferable to argue that the rural is no longer
'mappable' as a set of physical or social distinctions (or perhaps more
accurately, as we shall explore below, while the rural is clearly 'performed'
by mapping exercises these should be seen as only particular, partial and
incomplete versions of what the rural might be(come)). There is no essential
rural condition, no point of reference against which rurality can be
measured. Every practice of dividing and distinguishing the rural is satur-
ated with assumptions and presuppositions. It is, of course, impossible to
step outside these; the only alternative, we believe, is to adopt a reflexive
approach to rural studies, one that takes account of the ways in which we
do the dividing and the distinguishing, and that considers the way in which
our categories and concepts, the very accounts that we write, perform
power relations so that these might become more visible and contestable
(Murdoch and Pratt 1994).

A first step here is to admit that the Other has always lain at the margins
of our worlds. In constructing the rural we have often implicitly and
sometimes explicitly drawn a line across which only the most notable
figures in the landscape have been allowed to tread (e.g. Philo's powerful
groups). For this reason the rural is easily portrayed as a 'civilised retreat'
(Lowe et al. 1995), a zone where Sameness (British or English middle-class
whiteness and heterosexuality) is reasserted in the wake of a profound
postcolonial anxiety. The role that rurality plays in the process of Othering
is, therefore, reasonably clear; what remain unclear, however, are the experiences and identities of Otherness which might legitimately be encompassed
within a rural framework. We must push our analyses further, therefore,
as Philo suggests, into social realms that are occluded within the zone of
Sameness. But this move must be accompanied by some rethinking of how
we do what we do: too often 'to name is to possess, to domesticate is to
extend patronage. [For] we are usually only willing to recognise differences
so long as they remain within the domain of our language, our knowledge,
our control' (Chambers 1994: 30). The challenge is now to throw ourselves
into that ambiguous space 'in which differences are permitted a hearing,

in which both speakers and the syntax of conversation run the risk of modification' (ibid: 31). It is to embrace an 'ethics of difference', one which 'can express and encourage an openness of outlook based upon a freedom to move across border and boundaries in pursuit of new senses of self and other' (Pile and Thrift 1995: 21).

To summarise this section, we are claiming that whichever conceptions of space come to predominate will do so, as often as not, as a result of social struggles and the imposition of certain sets of power relations (see also Pratt 1991). However, once a web of meanings has been imposed and stabilised it will, in turn, give rise to a further set of effects; that is, ways of ordering the world, in which certain thoughts, statements, practices are deemed legitimate or illegitimate. As we have implied above, categories such as the rural, and the identities which are implicated in particular definitions, can be traced in terms of power effects. Thus, the concern with neglected Others in the rural domain points up the systems of classification which have dominated rural studies and how these have resulted in particular effects (i.e. neglect). This concern requires that we take a fresh look at our concepts and methods. Having considered some of the most general categories which guide academic work, in the next section we turn to elaborate in a little more detail some of the frames which have traditionally 'organised' analysis of the rural.

TOPOGRAPHIES OF THE RURAL: REGION, NETWORK AND FLUIDITY

It should by now be clear that the time for some modification in the conceptual and methodological tools that guide academic analysis is nigh. Our aim here is to take on board the challenges that confront us, as outlined above, while leaving scope for social scientific analysis of the rural. We are not sympathetic to the view that social science is inherently and irredeemably 'modernist' and, therefore, inappropriate to the study of such postmodern themes as difference and Otherness. We believe it is possible to be sociological and to be sensitive to diverse experiences and histories, and we believe sociology can still chart a course through the multiple spaces and territories that constitute the rural terrain. This sociology must, however, now recognise, as Marc Mormont puts it (1990: 34), that the rural 'is no longer one single space, but a multiplicity of social space . . . each of them having its own logic, its own institutions, as well as its own specific network of actors'. Thus there is no one unique and privileged vantage point, no one centre from which the rural can be captured and assessed. Rather, we must accept that academic studies of the rural will be characterised by partial, incomplete and contingent views of increasingly complex ruralities. Once we recognise this we must also acknowledge that we are no longer at the centre of the world. Our sense of the centre is being displaced. . . . The zone we now inhabit is open, full of gaps: an excess that is irreducible to a single centre or point of view. In these intervals . . . other stories, languages and identities can also be heard, encountered and experienced.

(Chambers 1994: 24)

The rural is contingent, fluid, detached from any necessary, stable socio-spatial reference point. Its meanings are asserted relationally (most notably in contradistinction to the urban) and are situationally specific; that is, we can know the rural only from and through particular socio-spatial positions. Moreover, taking Other ruralities seriously means that we must attend to the frameworks of organisation that allowed these so-called Others both to disappear from view and then to re-emerge into the light of academic attention. Therefore, in this section we consider three main ways of seeing rural space. First, we examine the boundaries that can encircle our spaces, directing attention to those who lie within but excluding those without. Second, we move on to the relationships in which our subjects are enmeshed and show that, while this concern allows for the examination of processes of interaction and relations of power, it ignores those who are not incorporated, those who lie outside dominant configurations of the rural. This leads us into a realm that lies in between inclusion and exclusion, inside and outside. In this third framework our rurals are open and fluid. The investigation of this in-between type of rurality requires modes of analysis which are flexible enough to follow these relatively fluid social spaces as they emerge, stabilise and fragment.

The terms we use to characterise these three ways of seeing are region, network and fluidity. We have adopted these from a study in the sociology of science (a forum which we have found particularly useful in the conduct of rural studies – see Murdoch and Pratt 1994; Murdoch and Marsden 1995) by Anne Marie Mol and John Law (1994). We recognise that our own use of these concepts may not accord with theirs. However, our intentions are comparable, for we also wish to explore the *topological suppositions* which frame the performance of difference' (Mol and Law 1994: 642; emphasis in the original). We are concerned to understand how topographies of the rural might be employed in ways which render sets of power relations transparent. And, again in common with Mol and Law, we believe our mode of analysis should make such an aim explicit and should be tailored to meet that requirement. It is in this spirit that we employ their terms.

In line with Mormont's conception of the rural, Mol and Law do not believe that the 'social' exists as a single spatial type. Rather, they argue,

it performs several *kinds of space* in which different 'operations' take place. First, there are *regions* in which objects are clustered together

58

and boundaries drawn around each cluster. Second, there are *networks* in which distance is a function of the relations between the elements and difference a matter of relational variety. . . . [However] sometimes, we suggest, neither boundaries nor relations mark the difference between one place and another. Instead, sometimes boundaries come and go, allow leakage or disappear altogether, while relations transform themselves without fracture. Sometimes, then, social space behaves like a fluid. [Fluid spaces] do without the solidity of regions and the formality of networks [emphasis added].

(Mol and Law 1994: 643)

We will briefly outline our understanding of each of these terms – region, network, fluidity – and will provide a preliminary assessment of their potential utility in rural studies.

Region

Of the three concepts, region is the most familiar to rural geographers. This concept has had a variable history within the (sub)discipline, being seen, on the one hand, as simply a physical concept – that is, a differentiated segment of earth space – and, on the other, as a social concept – that is, as a series of settings for social interaction (Thrift 1983). In line with our earlier comments on the relational nature of space within poststructuralism, Mol and Law see the region as an inherently social construct but one wherein

space is exclusive. Neat divisions, no overlap. Here or there, each space is located at one side of a boundary. It is thus that an 'inside' and an 'outside' are created. What is similar is close. What is different is elsewhere.

(Mol and Law 1994: 647)

This sense of the region has many resonances when we turn to consider the rural. In England, for instance, we need only think of the landscape, that great creation of the eighteenth century, to imagine how the rural comes to be a delimited, exclusive space. As Barrell (1992: 131) says of the representation of labourers within the paintings of Constable, 'the farmworkers are either invisible, their presence in the countryside to be inferred only from the effects of their labour, or else they are minute and distant, diminished into children, into specks of white in the far background'. This view of the countryside acutely illustrates how the rural can be captured and rendered exclusive through an individualistic yet dominant way of seeing, 'a way of seeing which separates subject and object, giving lordship to the eye of a single observer' (Cosgrove 1984: 262).

Such exclusive representations of rural space might seem, at first glance, a long way from regionalism within geography. Yet positivism imbued geography with the scientific urge to tame the landscape in a not dissimilar fashion. As Thrift (1994: 206) points out, for Vidal de la Blache (a pioneer of geographical analysis), geography was the science of landscape. He quotes Ross (1988), who argued that 'Videlian geography takes its model from the taxonomic dream of the natural sciences and is resolutely turned towards description; the geographer finds himself [sic] facing a landscape: the perceptible, visible aspect of space'. And as geography matures as a discipline so a variety of techniques are employed to render this landscape even more visible: Harley (1992: 231), in a discussion of cartography, believes that 'one effect of accelerated technological change – as manifest in digital cartography and geographical information systems – has been to strengthen its positivist assumptions'. This positivist way of seeing is summarised by Pile and Thrift (1995: 45) as a scopic regime which

> valorises the neutrality of seeing: the world is turned into a set of geometrical arrangements based on an abstract, fixed, universal, iso-tropic and material understanding of space; indeed, it is this 'space' which is properly presented in the generic map – a flat, supposedly all-seeing (if not all-showing) picture of (part of) the world.

The rural has been bound up into this science of the landscape. One need only think of the geographer L. Dudley Stamp and the way his inter-war land use survey influenced the ideas of urban containment which became so central to the post-war planning system (see Hall *et al.* 1973) to recognise the power of this form of representation. Even where the social characteristics of rurality have become pre-eminent, attempts to map rural space tend to reproduce the neat divisions, the insides and outsides. As Halfacree (1993: 23) notes, definitions of rurality, such as Cloke's rurality index (Cloke 1977; Cloke and Edwards 1986), tend to concentrate upon that which is observable and measurable. Such 'empiricism', he believes, 'accepts that the rural exists and concerns itself with the correct selection of parameters with which to define it'. It is tempting, therefore, to simply dismiss this rural geography of the region out of hand; to portray it as outmoded, irredeemably positivist or modernist, unsuited to the frac-tured and diverse countrysides that we now see before us. This would be a mistake for a number of reasons. First, the region is undoubtedly a powerful construction. The portrayal of space as an exclusive domain in which diverse entities are standardised and homogenised can be studied as a process of 'normalisation' (Harley 1992, after Foucault) which in itself gives rise to powerful representations of the world. In turn, these representations influence 'powerful' actors (e.g. planners) and the ways they choose to intervene in the world. Second, these techniques are also powerful in their own right; they inscribe what is visible and thus, in Latour's (1987) phrase, allow 'action at a distance' on entities far removed in space and time (using a map we can, in a very real sense, 'visit' a place without ever 'leaving home'). As Chambers (1994: 92) reminds us, 'with a map in our hands we can begin to grasp an outline, a shape, some sort of location'. There may be more to this location than can ever be mapped, yet the demarcation of regions does capture some elements even if it is just those that are the most visible. While such visions of the rural are always incomplete, and are never enough, they should be seen for what they are: sets of powerful inscriptions which 'perform' certain powerful ruralities.

Networks

Of course, by demarcating 'insides' regional ruralities alert us to 'outsides'; they shadow the 'landscapes of exclusion' that Dave Sibley (e.g. 1992) has long insisted remain intrinsic to the arrangement of rural (and urban) spaces. However, the main problem with regional rurals is that they not only obscure the fragmented nature of contemporary socio-spatial forma-tions, but leave opaque the processes of change that continually give rise to regional shapes. The processes of change, it is argued, do not respect the discrete boundaries that circumscribe the rural; they transgress such distinctions and travel over much longer distances and times. Thus spatially proximate clusters such as regions neglect the way in which relationships are forged over long distances. These relations, Mol and Law believe, are better thought of as 'networks' wherein the network brings together two or more locations that may be far away from each other on a regional map:

> In a network space . . . proximity isn't metric. And 'here' and 'there' are not objects or attributes that lie inside or outside a set of bound-aries. Proximity has, instead, to do with . . . the network elements and the way they hang together. Places with a similar set of elements and similar relations between them are close to one another, and those with different elements or relations are far apart.
>
> (Mol and Law 1994: 649)

Here the region is folded by the network configurations. Within the net-works, spaces are constructed as nodes form and links are imposed. Networks give rise to regional effects, as cross-cutting sets of relations issue outcomes that may coalesce in the form of 'places'. Yet to make sense of the key social relations that give rise to such regional effects the task becomes that of following the networks as actors are bound together, identities are forged and power relations stabilised.

In the rural domain this type of work is still at an early stage but has been utilised most notably in studies of the food system and the develop-ment process. In the former body of work the incorporation of agriculture

61

into the food system shows how local (agricultural) effects (such as changing farm structures, patterns of cropping, inputs and so on) result from sets of power relations established higher up the food chain (by, for instance, retailers) (see Fine 1994). This approach, as Whatmore (1994) explains, has taken studies of agriculture beyond the farm gate to the technological and economic interrelations between farming, agricultural input suppliers, scientific research and development, food processing, retailing and the regulatory activities of various state agencies. A vast web of interlinkages called the food chain has been uncovered. In the process, it has become clear that agriculture is tied into sets of 'vertical' relations in this chain which are far more important than 'horizontal' relations to other aspects of life in rural areas (Marsden *et al.* 1990). The rural cannot, therefore, be equated, in a regional sense, with the agricultural.

In the latter (horizontal) area, the processes giving rise to exclusive regional ruralities (e.g. the English middle-class, white, hetereosexist spaces we are all familiar with) have been analysed in work on rural development as networks, showing how local actors are tied into long chains of power relations (Murdoch and Marsden 1994, 1995). In the case of land development interests, for example, it is clear that change at the local level can be properly understood only when we chart the relations between the local agents of change and those situated much further away (at, for instance, the 'national' level). It has also become evident that patterns of development and conservation or preservation consistently emerge from forms of collective action (e.g. Cloke and Little 1990; Lowe, 1977; Murdoch and Marsden 1994; Short *et al.* 1986). In order to shape rural space – the protection or development of which acts as a spur to much political action – actors need to pool resources, build alliances and act in concert with others.

Effectively this work is a study of the powerful and shows in detail how these actors become powerful and sustain their power. The rural is configured by these networks (they provide in Massey's (1991) terms its 'power geometry'), and in part the analysis of these helps explain how landscapes become exclusive, how Others are excluded from the scene by powerful actors. The problem with this approach, however, is that it forces the social scientist to follow the network builders and, as Law (1991: 11) notes, 'it becomes *difficult to sustain any kind of critical distance from them*. We take on their categories. We see the world through their eyes. We take on the point of view of those whom we are studying' (emphasis in the original). Moreover, it is only powerful actors who tend to get followed. While the network approach is good at showing the contingency of power relations by documenting in detail how the powerful become powerful it tells us nothing about those who lie outside the (power) networks. Those who lack resources, a voice, visibility, will continue to be neglected if we simply concentrate on powerful networks. Although network analysis helps in understanding how the rural becomes an exclusive, homogeneous terrain it does not direct us towards those who fall into the gaps between the networks. It does not enable us to hear Other voices, Other experiences or to understand Other rurals.

Fluidity

We are driven again, therefore, to ask whether there are other spaces around, 'spaces that have topographical properties which aren't like those of regions and networks' (Mol and Law 1994: 653). We have already stressed our willingness to conceptualise the rural as a fluid malleable space, one comprising heterogeneous flows and complexities. Thus we now confront a third space, a fluid space, in which Mol and Law (ibid) believe that

> there are often no clear *boundaries*. Typically the objects generated inside them – the objects that generate them – aren't well defined. Thus, even the boundary between the normal and pathological . . . isn't given once and for all. Any attempt to fix it tends to falter [original emphasis].

In this fluid space nothing comes neatly packaged into insides and outsides, Sames and Others, here and there. Fluidity ensures that there are varying shades and colours, that things are much more mixed up:

> A fluid space . . . isn't quite like a regional one. Difference inside a fluid space isn't necessarily marked by boundaries. It isn't always sharp. It moves. And a fluid space isn't quite like a network, either. For in a fluid elements inform each other. But the way they do so may continually alter. The bonds within fluid spaces aren't stable. Any single component – if it can be singled out – can be missed.
>
> (Mol and Law 1994: 663)

It is fair to say that this fluid space is now gripping the social science imagination. Postcolonial writers, such as Homi Bhabha (1990), have suggested the notion of a 'third space' as an in-between, fluid, intertextual space. It is the acknowledgement of this 'third', 'ambiguous' or 'hybrid' space which has brought to light the Other and forced a reassessment of the Same, ensuring that no sharp boundaries between Us and Them can be maintained. Now static binary divisions give way to fluid, incomplete, open forms of identity: as Chambers (1994: 82) says,

> to talk of differences, even radical and incommensurable ones, in economic, political and cultural terms, and of their embodiment in ethnicity, gender and sexuality, is to talk of an understanding of the making of identities in movement, under and in, processes.

There are now no fixed points of reference, no privileged points of view; simply a swirling, viscous, partially stable, partially enclosed, movement of social entities.

In rural studies the call to attend to neglected Others forces a recognition of this third, fluid space. As yet, however, work of this kind is still thin on the ground (a situation to be partially rectified by the present volume), but a few studies which emanate from the earlier traditions of rural studies do give some sense of what a third rural space might look like. We are thinking here of anthropological work on rural communities. While much of this work has traditionally been the object of criticism, notably for its representation of these social forms as overly consensual and stultifyingly harmonious (the classic critique is Bell and Newby (1971)), the anthropological method of long-term residence in the community has often led to great sensitivity to the diversity of communal identities. Fiona Bowie (1993: 168–9), for instance, makes the following comment on the virtues of an anthropological approach in the context of Wales:

> Wales presents to the rest of the world a coherent picture of cultural self-sufficiency and a firm sense of identity. What outsiders see, however, is not so much Wales as their own reflection, or stereotypes of Welshness.... As one begins to penetrate beyond this refracted image of Welshness, not least by learning the Welsh language, the unproblematic and monolithic nature of Welsh identity begins to fragment. One is left not so much with a coherent notion of Welshness ... as with a sense of many conflicting and interlocking definitions of identity which actively compete for symbolic space and public recognition.

It is in the course of this 'penetration' of rural cultures that in-between spaces emerge. This is strikingly evident in Nigel Rapport's (1993) study of the village of Wanet in Cumbria, which shows an acute sensitivity to 'the seeming sameness' of such analytical categories as 'rural', 'community', 'kinship', 'class', social structure' and so forth. In a rural community such as 'Wanet', Rapport (1993: 41) believes,

> we might find that rather than building blocks which always cause the replication of one kind of collective structure (an overarching, equilibrial community) items of routine behaviour serve as ambiguous and malleable forms which may crop up in very different circumstances, be combined in different ways, and, to their protagonists, mean different things.

Despite his identification of many communal forms of behaviour, Rapport finds no standard definition of what these entail among the inhabitants of 'Wanet': in use, the forms which many would agree upon as common and proper come to be mediated by a diversity of individual ends. Any one behavioural form can, in specific situations, have a number of different meanings:

> behavioural commonalties are personalised in usage and come to be animated in possibly idiosyncratic fashions. They become instruments of diversity and difference, and yet the conditions of their use remain essentially public, and it is in co-ordination with significant others and in certain routine and limited ways that these meanings come to be made.
>
> (Rapport 1993: 170)

The distinction between individual difference and communal similarity begins to be blurred in this analysis. In Rapport's 'Wanet' the community works only by maintaining a balance between the idiosyncrasy of personal expression and the reproduction of routine behavioural norms. He thus points to a 'tension' between what is shared or exchanged and the idiosyncrasies of creative expression:

> The beauty of the exchange for me is this awful tension between the surface exchange, the orderly conversational form, the shared knowledge of interactional systematics – how to speak, when, and in what manner – and the unique visions, the limitless avenues of thought, the wild disorders of contradiction that can be motivating the exchange, causing its regular reoccurrence, and dancing delightedly but invisibly around its expression.
>
> (Rapport 1993: 163–4)

A third space begins to emerge, therefore: it is a gap between the accepted (regional) conception of a homogeneous community and stable selves; it is a fluid, ambiguous space full of the complexities of norm and difference, forever changing in situation after situation.

Rapport successfully renders this fluid rurality using the language of symbolic interactionism, a mode of analysis which is well suited to situationally grounded research. However, it is also important to recognise the role of methodology in all this. Rapport's experience of 'fitting in' shapes his analysis in profound ways. Much of Rapport's own experience of 'Wanet' is a sustained attempt to strike a balance between normal community behaviour and self-expression. By immersing himself in the rural, while opening himself up to the experience of others, insights into a 'third' space are achieved. There is much to be recommended in this type of social science enquiry, for in some ways it forces the analyst to enter into an experience of Otherness. By becoming a 'stranger' in the rural, by coming from elsewhere, from 'there' and not 'here', and hence by being both 'inside' and 'outside' the situations at hand, we can begin to experience that estrangement, that 'uncanny displacement' (Chambers 1994: 6), which can so often characterise the experience of Otherness. In other words, we are forced to confront strange ruralities.

To summarise this section, then, we have identified three types of rural space: regional space, network space and fluid space. It should be noted that the portrayal of these spaces has included a mixture of concepts or ways of seeing and that which is seen: the rural(s). The distinction between ways of seeing and that which is seen is difficult if not impossible to draw and we have not attempted to do so. We have no wish to argue that these three spaces can be considered in developmental terms; that is, we are progressing from a regional space to a fluid space. On the contrary, along with Mol and Law (1994: 663), we want to emphasise that the 'three topologies have *intricate relations*. They co-exist.' They perform the rural in their different ways and the effects are all around us. Social science is implicated in these performances, for it plays its role in conjuring up certain rurals and neglecting others. We have proposed these three rural spaces in order to allow more balanced accounts of diverse and complex rurals to be formulated. The three concepts are offered here in the hope that they enable rural studies to capture this diversity and complexity in terms which allow Other experiences and voices to be heard.

CONCLUSION

We began this chapter by acknowledging that rural studies has traditionally tended to focus upon the powerful in rural areas and has been concerned with understanding the rural in terms of modernisation and rationalisation. We can discern shifting research priorities leading to increasing attention on division and difference in the countryside. This movement culminates in Chris Philo's plea for rural studies to engage with previously neglected Others, for the subdiscipline to understand that there is far more within the rural terrain than has previously been imagined. Our own contribution to this shift in focus has been to ask that we are careful about our conceptual and methodological approaches to these new rurals. In order to facilitate a careful reconsideration of academic work in this area we have provided three 'middle-level' concepts which might fruitfully organise future discussions of the increasingly strange ruralities that are likely to confront us. *Region* is proposed as emblematic of traditional approaches to the rural and as such stands for the purified representations of rural space that have characterised both rural studies and other rural discourses. The term *network* captures many current research concerns which seek to illuminate the activities of the powerful as they order the rural and marginalise Others. *Fluidity* is introduced to clarify the nature of the 'third' space which contemporary poststructuralist writers have marked out as their chosen terrain. Here identities are only provisionally stabilised, boundaries only provisionally marked and rurality rendered increasingly strange. For the critical social scientist this third, fluid space is currently the most challenging aspect of rurality.

While these concepts have been introduced to allow rural studies to represent the rural more fairly – giving voice to the previously neglected, for instance – they should also mark out the limits of what social science can and should do: no longer can one framework render the whole rural world for all to see. We should, therefore, in the spirit of reflexivity, aspire to modesty in our endeavours. To emphasise this point we close with a quote by John Law (taken from Callon and Law 1995: 504) which attempts to mark out the legitimate scope of our ambitions in relation to the neglected Others that will soon be crowding onto the rural studies research agenda:

> To imagine that we can assimilate the Other in any of its forms is hubris. Instead, it seems to me that these Others will ignore us for most of the time. Instead, they will continue, as they always have, to perform their specific forms of agency to one another. And all that we can do is to say that these performances go on. And then to create appropriately monstrous ways of representing them on those rare occasions when our paths happen to cross and we find, for a moment, that we need to interact with them.

REFERENCES

Ambrose, P. (1975) *The Quiet Revolution*, Methuen, London.

Barrell, J. (1992) 'Sportive labour: the farmworker in eighteenth-century poetry and painting', in Short, B. (ed.) *The English Rural Community: Image and Analysis*, Cambridge University Press, Cambridge.

Bell, C. and Newby, H. (1971) *Community Studies*, Allen and Unwin, London.

Bowie, F. (1993) 'Wales from within: conflicting interpretations of Welsh identity', in MacDonald, S. (ed.) *Inside European Identities*, Berg, Oxford.

Callon, M. and Law, J. (1995) 'Agency and the hydrid collectif', *South Atlantic Quarterly* 94(2), 481–507.

Chambers, I. (1994) *Migrancy, Culture, Identity*, Routledge, London.

Cloke, P. (1977) 'An index of rurality for England and Wales', *Regional Studies* 11, 31–46.

Cloke, P. and Edwards, G. (1986) 'Rurality in England and Wales 1981: a replication of the 1971 index', *Regional Studies* 20, 289–306.

Cloke, P. and Little, J. (1990) *The Rural State*, Clarendon, Oxford.

Cloke, P., Philo, C. and Sadler, D. (1992) *Approaching Human Geography*, PCP, London.

Cloke, P., Doel, M., Matless, D., Phillips, M. and Thrift, N. (1994) *Writing the Rural: Five Cultural Geographies*, PCP, London.

Cosgrove, D. (1984) *Social Formation and Symbolic and Lanscape*, Croom Helm, London.

Department of the Environment (1995) *Rural England*, HMSO, London.

Fine, B. (1994) 'Towards a political economy of food', *Review of International Political Economy* 1(3), 519–46.

Foucault, M (1982) 'The subject and power', in Dreyfus, H. and Rabinow, P. (eds) *Michael Foucault: Beyond Structuralism and Hermeneutics*, University of Chicago Press, Chicago.

Gupta, A. and Ferguson, J. (1992) 'Beyond culture: space, identity and the politics of difference', *Cultural Anthropology* 7, 6–23.

Halfacree, K. (1993) 'Locality and social representation: space, discourse and alternative definitions of the rural', *Journal of Rural Studies* 9(1), 23–37.

—— (1995) 'Talking about rurality: social representations of the rural as expressed by research in six English parishes', *Journal of Rural Studies* 11, 1–20.

—— (1996) 'Our place in the country: travellers and the "rural idyll" ' *Antipode* 28(1), 42–72.

Hall, P., Gracey, H., Drewett, R. and Thomas, R. (1973) *The Containment of Urban England*, Allen and Unwin, London.

Harley, J. (1992) 'Deconstructing the map', in Barnes, T. and Duncan, J. (eds) *Writing Worlds: Discourse, Text and Metaphors in the Representation of Landscape*, Routledge, London.

Latour, B. (1987) *Science in Action*, Open University Press, Milton Keynes.

Law, J. (1991) 'Introduction: monsters, machines and socio-technical relations', in Law, J. (ed.) *A Sociology of Monsters: Essays on Power, Technology and Domination*, Routledge, London.

Lowe, P. (1977) 'Access and amenity: a review of local environmental pressure groups in Britain', *Environment and Planning A* 9, 35–58.

Lowe, P., Murdoch, J. and Cox, G. (1995) 'A civilised retreat? Anti-urbanism, rurality and the making of an Anglo-centric culture', in Healey, P., Cameron, S., Davoudi, S., Graham, S. and Madani-Pour, A. (eds) *Managing Cities: The New Urban Context*, Wiley, London.

Marsden, T., Lowe, P. and Whatmore, S. (1990) 'Introduction', in Marsden, T., Lowe, P. and Whatmore, S. (eds) *Rural Restructuring: Global Processes and Their Responses*, Fulton, London.

Massey, D. (1991) 'A global sense of place', *Marxism Today*, June, 24–9.

Mol, A. and Law, J. (1994) 'Regions, networks and fluids: anaemia and social topology', *Social Studies of Science* 24, 641–71.

Mormont, M. (1990) 'Who is rural? Or, how to be rural', in Marsden, T., Lowe, P. and Whatmore, S. (eds) *Rural Restructuring: Global Processes and Their Local Responses*, Fulton, London.

Murdoch, J. (1995) 'Middle-class territory: some remarks on the use of class analysis in rural studies', *Environment and Planning A* 27, 1213–30.

Murdoch, J. and Marsden, T. (1994) *Reconstituting Rurality: Class, Community and Power in the Development Process*, UCL Press, London.

—— (1995) 'The spatialisation of politics: local and national actor-spaces in environmental conflict', *Transactions of the Institute of British Geographers* NS 20, 368–80.

Murdoch, J. and Pratt, A. C. (1993) 'Modernism, postmodernism and the post-rural', *Journal of Rural Studies* 9, 411–27.

—— (1994) 'Rural studies of power and the power of rural studies', *Journal of Rural Studies* 10, 83–7.

Newby, H. (1980) *Green and Pleasant Land? Social Change in Rural England*, Hutchinson, London.

Newby, H., Bell, C., Rose, D. and Saunders, P. (1978) *Property, Paternalism and Power*, Hutchinson, London.

Pahl, R. (1970) *Readings in Urban Sociology*, Pergamon, Oxford.

Philo, C. (1992) 'Neglected rural geographies: a review', *Journal of Rural Studies* 8, 193–207.

—— (1993) 'Postmodern rural geography? A reply to Murdoch and Pratt', *Journal of Rural Studies* 9, 427–36.

Pile, S. and Thrift, N. (1995) 'Mapping the subject', in Pile, S. and Thrift, N. (eds) *Mapping the Subject*, Routledge, London.

Pratt, A. C. (1991) 'Discourses of locality', *Environment and Planning A* 23, 257–66.

—— (1996a) 'Rurality: loose talk or social struggle?' *Journal of Rural Studies* 12 (1), 69–78.

—— (1996b) 'Deconstructing and reconstructing rural geographies', *Ecumene* 3 (3), 345–50.

Rabinow, P. (ed.) (1986) *The Foucault Reader*, Penguin, Harmondsworth.

Rapport, N. (1993) *Diverse World Views in an English Village*, Edinburgh University Press, Edinburgh.

Short, J., Fleming, S. and Witt, S. (1986) *House Building, Planning and Community Action*, Routledge and Kegan Paul, London.

Sibley, D. (1988) 'Purification of space', *Environment and Planning D: Society and Space* 4, 409–21.

—— (1992) 'Outsiders in society and space', in Anderson, K. and Gale, F. (eds) *Inventing Places: Studies in Cultural Geography*, Longman/Wiley, London.

Thrift, N. (1983), 'On the determination of social action in space and time', *Environment and Planning D: Society and Space* 1, 23–57.

—— (1989) 'Images of social change' pp. 12–42, in McDowell, L., Sarre, P. and Hamnett, C. (eds) *The Changing Social Structure of Britain*, Sage, London.

—— (1994) 'Taking aim at the heart of the region', in Gregory, D., Martin, R. and Smith, G. (eds) *Human Geography: Society, Space and Social Science*, Macmillan, London.

Whatmore, S. (1994) 'Global agro-food complexes and the re-fashioning of rural Europe', in Amin, A and Thrift, N. (eds) *Holding Down the Global*, Oxford University Press, Oxford.

Wirth, L. (1938) 'Urbanism as a way of life', *American Journal of Sociology*, 44, 1–24.

4

CONTRASTING ROLES FOR THE POST-PRODUCTIVIST COUNTRYSIDE

A postmodern perspective on counterurbanisation

Keith Halfacree

[E]ven as we acknowledge ourselves to be de-centred and fragmented subjectivities, the gendered constructs of patriarchy and mouthpieces of a discursive ventriloquism, we also seem to rediscover a centre, the existential, angst-ridden self who must also make sense of it, and seek to reorganise desire, re-read the world, adjust behaviour, and so on, in the light of that awareness.

(Soper 1993: 25)

INTRODUCTION

Migration from an urban to a more rural residential environment is commonplace in most highly developed countries. In Britain, such migration takes place with a backdrop of an agricultural industry mired in a state of crisis. Indeed, academic analysis of the depth and prolonged character of this crisis have led to suggestions that we are witnessing a shift from a 'productivist' to a 'post-productivist' era in the countryside as a whole. With such a shift comes the opening up of a space for relatively novel actors to stamp their identity upon the British countryside.

This chapter attempts to provide some pointers to the contribution that urban–rural migrants may be making to the creation of a post-productivist countryside through examining rival theoretical interpretations of the motivation and attitudes of the migrants. Counterurbanisation is examined in terms of its premodern, modern and postmodern characteristics, as shaped by and through the migrants' increasingly postmodern experiences of everyday life. It is argued that while some aspects of counterurbanisation can be seen as reactionary and others as an 'irresponsible' celebration of the times, the migration trend also contains a strong reflexive strand,

70

engaging critically with the existential void of our postmodern existence. From the latter perspective, which the chapter argues is unstable and by no means coherent or universal, migrants have the potential to stir a radical ingredient into the emerging post-productivist countryside. Some signs of this are shown to be apparent in recent environmental debates over the fate of the British countryside. First, however, the contours of the shift from a productivist to a post-productivist countryside must be mapped out.

THE EMERGENCE OF THE POST-PRODUCTIVIST COUNTRYSIDE

For British agriculture, the years defined as productivist lasted from around the end of the Second World War and the rebuilding of agriculture in the light of the Agriculture Act of 1947 to around the late 1970s. Although driven by the often harsh demands of the 'agricultural treadmill' (Ward 1993), the productivist era provided a sense of security to members of the agricultural community with respect to their land rights, land use, finance, politics and ideology, bolstered by a highly corporatist relationship between the agricultural industry and the British government. In short, agricultural predominance in the countryside was regarded as benign, with agriculture promoted as a progressive and expanding food production-oriented industry (Marsden *et al.* 1993). Moreover, other institutions concerned with the British countryside tended to accept and respect the role that was accorded to agriculture. For example, the 'agricultural exceptionalism' (Newby 1987: 216) of the Scott Report of 1942, whereby agriculture was virtually exempted from the planning controls imposed on other forms of industry, sent a clear message that agriculture's hegemonic position in the countryside was both appropriate and secure.

Although productivism was not beneficial to all farmers throughout its years of hegemony, particularly with respect to those who were unable to accede to the demands of the agricultural treadmill, it was not until the 1970s that it faced sustained attack. There is not the space here to detail the threats which the productivist regime encountered; suffice it to note Marsden *et al.*'s (1993) emphasis on the economic and other contradictions in the 'Atlanticist' food order, the post-*Silent Spring* concern with the negative environmental effects of many agricultural activities, and the impacts of service class in-migration to rural areas (Cloke and Goodwin 1992: 327–8). As a result of these contradictions, British farmers began both to feel and, indeed, to experience an increasing sense of insecurity and uncertainty as regards their position in both agriculture and rural life generally. Moreover, this was also the perception of the general public, who increasingly questioned and criticised farmers' status as guardians of the countryside and, with an eye to their own financial positions in a time of economic recession, the monetary and other support that giving farmers this position entailed. Consequently, there emerged high levels of debt and depression within the farming community and a growing involvement in non-food-producing activities through diversifying into such activities as bed and breakfast accommodation.

The ongoing crisis in productivism is not just of significance to the agricultural community as it also signals a crisis for all those implicated in the productivist era. Overall, the crisis may lead to the suggestion that the hegemonic domination of rural areas and rural society by agriculture, which, of course, extended further back than 1945, is itself coming to an end. Technologically, such a shift is increasingly possible given advances in intensification, biotechnology, etc. Hence, post-productivism may signal a search for a new way of understanding and structuring the countryside. A space in the imagination is opening, whereby non-agricultural interests and actors are given an opening to strive to create a rurality in their image. This space has been bolstered in a more 'physical' sense by an increased recognition of the presence of supposedly 'surplus' land. Consequently, Cloke and Goodwin (1992) suggest that previous local 'structured coherences' underpinned by agricultural relations are being replaced by a more fragmented pattern reflecting the diverse ways in which rural space is currently being commodified.

COUNTERURBANISATION AND NEW SPATIAL DIVISIONS OF LABOUR

The turn towards a post-productivist and in many ways a post-agricultural future for the countryside is most clearly represented in the importance of counterurbanisation and counterurbanisers in the production of contemporary rural space. Counterurbanisation (Champion 1989, 1992), in brief, can be defined as the population revival and growth, via net migration, of 'rural areas', together with the corresponding population decline of the cities and large towns (Fielding 1982). It reflects both the increasing use of rural space for non-agricultural purposes and the predominance of consumption interests over production interests, with the rural as a space of residence. Investigation of the population turnaround reveals a complex pattern of causality, with a clear absence of any single explanatory process. Some of the suggested causes of counterurbanisation are outlined in Table 4.1. Following Moseley (1984), we can summarise these accounts in terms of two major emphases: job-led accounts and people-led accounts.

Job-led accounts of counterurbanisation generally link this migration trend to the emergence of a new spatial division of labour (e.g., Fielding 1982), where the countryside and, in particular, the towns within it are net beneficiaries of both manufacturing and service employment at the expense of the large towns and cities. This process incorporates the so-called 'rural-

Table 4.1 Explanations for counterurbanisation given in the literature

1 The expansion of commuting fields round employment centres
2 The emergence of scale diseconomies and social problems in large cities
3 The concentration of rural population into local urban centres
4 The reduction in the stock of potential out-migrants living in rural areas
5 The availability of government subsidies for rural activities
6 The growth of employment in particular localised industries such as mining, defence and tourism
7 The restructuring of manufacturing industry and the associated growth of branch plants
8 Improvements in transport and communications technology
9 The improvement of education, health and other infrastructure in rural areas
10 The growth of employment in the public sector and personal services
11 The success of explicitly spatial government policies
12 The growth of state welfare payments, private pensions and other benefits
13 The acceleration of retirement migration
14 The change in residential preferences of working-age people and entrepreneurs
15 Changes in age structure and household size and composition
16 The effect of economic recession on rural–urban and return migration
17 The first round in a new cyclic pattern of capital investment in property and business

Source: Champion (1989: 236–7)

isation' of industry (Fothergill *et al.* 1985), familiar from the work of Fothergill and Gudgin (1982) onwards. Moreover, as Townsend (1993) has recently demonstrated, in spite of a number of 'hiccups' due to economic recession, this urban–rural shift in British economic geography does not appear to be a passing phase. Hence, the migration of people to more rural areas reflects their need to obtain employment, while the new job opportunities presented in rural areas also work to reduce out-migration.

From such a perspective, Harvey (1989: 340) is able to align counterurbanisation with what he has termed 'flexible postmodernity', while the metropolis is associated with 'Fordist modernity'. Indeed, we may wish to go further than Harvey and associate the population turnaround with the emergence of a 'post-industrial society' (Bell 1973), especially if we emphasise the importance of new service employment in rural areas – from tourism to teleworking – rather than rural manufacturing. While such a linking of counterurbanisation with a post-Fordist spatial division of labour will undoubtedly prove at least as problematic as the very concept of post-Fordism itself (Pollert 1988; Sayer 1989), such an economic understanding appears valuable. As regards the reasons for this ruralisation of industry, Townsend (1993) summarises the current level of understanding. He stresses the continued significance of the space and other physical constraints posed by urban locations, which were emphasised by Fothergill and Gudgin (1982). In addition, he mentions the importance of regional labour pressures and the lower wage rates characteristic of rural locations, which provide a further incentive to potential employers. However, he also mentions the often recognised but rarely emphasised 'other psychological and social class attractions of the "rural life" ' (Townsend 1993: 219). When we consider these attractions directly (for example, Halfacree 1994), we discover their central importance to the counterurbanisation trend. This necessitates attention being paid to people-led accounts of counter-urbanisation.

COUNTERURBANISATION AND THE CONTEXT OF EVERYDAY EXPERIENCE[1]

People-led explanations

People-led explanations, as the phrase implies, regard counterurbanisation very much as a spatial act which individuals freely enter into, whether as single persons or as part of a household. In other words, people are 'voting with their feet' in their migrations from urban to rural locations. When exploring counterurbanisation using this perspective, attention must be given to the highly positive 'appeal' of the rural residential environment to many migrants, as the environment is unlikely to be incidental to their moves. As Cloke (1994: 164) observes from his own experience, 'people do make decisions . . . which presume a category rural'. Elsewhere (Halfacree 1996b), I have defined this appeal as 'ruralism'.

A strong desire to live in a more rural residential environment is consistently demonstrated in British surveys. As one example from many, a Gallup poll in 1989 found that 72 per cent of British adults said that they would rather live in the country than in the city (King 1989). This figure had increased from 61 per cent in 1939. The 'pleasures' of the countryside mentioned by the Gallup respondents are listed in Table 4.2. While some of them reflect leisure interests, such as fishing, the overwhelming majority suggest a rural appeal which merits further investigation, such as the desire for space and solitude. Elsewhere, Sherwood (1984: 536ff.), in south Northamptonshire, investigated why a rural environment was preferred to an urban environment. The resulting attributes are listed in Table 4.3 and show many similarities with the Gallup survey. A need to investigate further the reasons why people appear to move to rural residential locations, assuming that the types of factors listed by King and Sherwood are important in this decision (Halfacree 1994), becomes especially apparent if we reject either any naturalisation of the rural appeal or any reduction of that appeal to market-based manipulation.

A naturalistic explanation of counterurbanisation comes from the argument that people have an innate human need to live in a rural residential environment. Thus, with improvements in personal mobility and wealth, all that has taken place is that people are now able to satisfy their needs

Table 4.2 The 'pleasures of the countryside'

Characteristic	Percentage citing
Countryside is more peaceful	71
Like feeling of space	54
Like trees and forests	50
Like flowers and wildlife	48
Countryside more natural	46
There aren't as many people	42
Like mountains	29
The country hasn't changed as much as towns and cities	24
Like feeling in touch with the past	19
Like farms and farming	16
Like fishing	11
Like hunting and shooting	3

Source: King (1989), from a Gallup survey

more completely. While accounts of counterurbanisation based upon the assumption that people have a natural urge to live 'closer to nature' are commonplace in popular discourse, they are usually more implicit in academic work. Nonetheless, Radford could claim that:

> a vast increase in motor-car ownership, coupled with acute urban land shortages and rising land prices, have led to a general exodus in the direction of clean air, spaciousness, and the prospect of living in or near a small-scale community.
>
> (Radford 1970: 3)

Whilst this explanation may be true, Radford provided little evidence to support her claim of the importance attached to rural living, implicitly naturalising the exodus. Similarly, Gracey (1973) assumed a 'search for Arcadia' and Connell (1978: ix) stated 'the almost universal desire to live among the fields and trees and recapture something of an idealised rural tranquillity'.

The major problem with the naturalistic explanation of counterurbanisation is, of course, that it neglects the way in which human needs and human nature are *produced* and 'are not spontaneously self-generating in the breasts of individuals' (Soper 1981: 9). In this respect, 'abstract universal needs' are a conceptual abstraction only, since supposedly universal needs are always 'concretely instantiated' within a specific societal context (Dickens 1989; Sahlins 1976; Smith 1984). Thus, we must explore the context of everyday life in which specific needs are formulated if we are to avoid an unhelpful reductionism.

We might instead argue that a desire for a 'house in the country' is a product of the capitalist marketplace. For example, rural houses tend to fetch premium prices and provide a useful additional rung for the housing ladder, which is highly beneficial for driving that market's capital accumu-

Table 4.3 Reasons given for preferring a rural residential environment

Reason	Percentage citing given reason
Like the country	30
Quietness	22
Lived in the village	18
Friendliness	12
Suitability for raising a family	5
Smallness	5
'Other'	8

Source: Sherwood (1984)

lation process. Moreover, the extent to which rural imagery is used in advertising makes plain its status and high-quality character (Burgess 1982; Thrift 1987; Williamson 1978). However, while serving as a useful critique of the naturalistic explanation, emphasis on market manipulation goes too far in granting the market deterministic power. As Bourdieu (1984: 231) notes, while the development of cultural taste necessitates both the production of the material goods implicated in the taste and the ability of consumers to adopt such a disposition, these 'economic' features are insufficient for this development. Taste and production are mutually reinforcing. Thus, we must seek among our everyday societal experiences for the basis of the ruralist sentiments latched onto by the market.

Critique of both naturalism and market manipulation means that we must be careful how we specify the people-led emphasis given to counterurbanisation. While the critique of both supports a clear role for human agency in driving the migration process, the emphasis upon societal context suggests that there is a danger in providing over-voluntarist accounts if we neglect this context and solely consider the proximate reasons given by the migrants for their relocations. Hence the concern of authors such as Fielding (1982) or Williams (1981) with people-led accounts of counterurbanisation.

The postmodern condition

Paralleling the debate as to whether the countryside is moving from productivism to post-productivism is the debate within academia more generally concerning whether we are moving from a modernist to a postmodernist era. It has now become something of a truism that the definition of postmodernism varies so much that it is itself postmodern! Therefore, the remainder of this chapter adopts the understanding of the concept espoused by Bauman (1992) and Smart (1993). They regard postmodernism primarily as a cultural and existential phenomenon, representing the growing realisation that the Enlightenment belief in inevitable-progress-through-science-and-reason is an unrealistic pipe dream. Thus, 'the promise of modernity to deliver order, certainty and security will remain unfulfilled' (Smart 1993: 27), with a consequent loss of a vision of the future (Mulgan 1989). In essence, postmodernism represents the apogee of modernism where the critical faculty so central to the Enlightenment project finally turns upon itself (also Cooke 1988; Giddens 1990, 1991). The dissipation of this auto-critique into everyday life generates the existential dilemmas of the resulting 'postmodern experience'.

Harvey (1989) characterises the immediate roots of the postmodern experience in terms of space–time compression, driven by the relentless demands of the capitalist system for capital acceleration and innovation. Temporally, 'flexible accumulation' relies upon accelerating the turnover time of capital, accentuating the volatility and ephemerality of fashions, commodities, the production process, norms, and even values and ideas (Harvey 1989: 285; also Jameson 1984). Spatially, communications systems, especially, symbolise the increasing 'annihilation of space through time' (Harvey 1989: 293; Zukin 1988). Consequently,

> the new space involves the suppression of distance . . . and the relent-less saturation of any remaining voids and empty places, to the point where the postmodern body . . . is now exposed to a perpetual barrage of immediacy from which all sheltering layers and intervening mediations have been removed
>
> (Jameson 1988: 351)

The power to ascribe the social meaning of space is removed from the people in those spaces and acquired by 'distant forces'. Spaces are reduced to 'flows and channels' (Castells 1983); they become 'phantasmagoric' (Giddens 1990: 108), as the local is shot through by the global (Smart 1993: 146). Thus, for Thrift (1994: 212, 227), the dominant 'structure of feeling' in the late twentieth century is one of 'mobility', synthesising speed, light and power into an essentially dynamic ontology. It is from within such a structure of feeling that the modernist sense of progress loses its way.

Other societal experiences further undermine our position in space and time (Smart 1993: 102). Changes in the class structure have fragmented existing working-class communities and resulted in the burgeoning of the service class (Lash and Urry 1987); the influence of electronic mass media has partially democratised knowledge but has also provoked a questioning of our selves and our institutions (Lash and Urry 1987; Meyrowitz 1985); while other societal changes present possible causes of further insecurity, such as the 'de-Oedipalisation of family life' (Pfeil 1988). Finally, con-suming signs divorced from fixed use values, as in Baudrillard's (1983a, 1988) 'television culture', provokes 'a nostalgia for the real' (Kroker 1985: 80).

Our psychological responses to instantaneity and disposability have long been recognised through analysis of the experience of 'modernity' (e.g. Gerth and Wright Mills 1970; Simmel 1971, 1978; Toffler 1970). However, these responses have been intensified considerably by the postmodern obliteration of progress and teleology. In order that psychologically rooted existential needs for identity and security (Eyles and Evans 1987; Levi and Andersson 1975) can be realised, we have to develop a sense of place (Pocock and Hudson 1978; Sack 1988), where time and space are brought together. Such integration is becoming increasingly problematic to achieve, as place is 'compromised' (Thrift 1994: 222). In summary, contemporary life is characterised by a crisis of 'ontological security', a concept referring to

> the confidence that most human beings have in the continuity of their self-identity and in the constancy of the surrounding social and material environments of action. A sense of the reliability of persons and things, so central to the notion of trust, is basic to feelings of ontological security; hence the two are psychologically closely related.
>
> (Giddens 1990: 92)

To return to counterurbanisation, while this migration trend may be post-Fordist or even post-industrial, it is not necessarily postmodern in itself; the spatial trend does not have to map the cultural experience. Indeed, the key question asked in the remainder of this chapter concerns the extent to which counterurbanisation and the motivations which underpin it, notably ruralism, are in tune with the postmodern existence, and the extent to which they are better seen as a reaction against, or a response to, that existence. In order to facilitate this investigation, we can explore counterurbanisation in the context of four strategies, suggested by Harvey (1989: 350–2; also Giddens 1990: 134–7; Smart 1993: 92), which can be adopted in the face of 'the fragility of ontological security in the wasteland of everyday life' (Giddens 1981: 193–4).

COUNTERURBANISATION IN A POSTMODERN CONTEXT: 1. PASSIVITY, CELEBRATION OR REACTION?

Passivity

An initial interpretation of counterurbanisation can immediately be dismissed. This is to regard it as a passive reaction, whereby the postmodern condition overwhelms the individual and reduces him or her to a 'stunned silence'. Such a response is reflected in many works of postmodern fiction and in much work on deconstruction. It is praised by Baudrillard (1983b) as the 'silent majority's' subversion of imposed cultural meanings. Counterurbanisation, however, is much too active to be regarded in this way, with the migrants themselves not conforming at all to Baudrillard's disinterested television 'channel hoppers'. Indeed, the very lack of 'silence' on the part of the counterurbanisers is apparent in any review of, *inter alia*, rural pressure groups (Lowe and Goyder 1983; Murdoch and Marsden 1994; Short *et al.* 1987). Moreover, the strength of pro-rural attitudes within 'British' culture (Potts 1989; Samuel 1989; Short 1991) fits poorly with any idea of cultural subversion. Instead, counterurbanisation appears as a very *active* response to the postmodern condition and appears to follow a narrative whose coherence has not been deconstructed to 'a rubble of signifiers' (Harvey 1989: 350).

Postmodern celebration

In contrast to passivity, counterurbanisation can be regarded as a much more active postmodern response. Such a response involves revelling in the diversity and ephemerality presented by the postmodern condition, in the manner advocated by Baudrillard (1988) and Lyotard (1984) in their rejection of meaning, reason or order. In short, it involves the abandonment of modernity (Smart 1993: 92) and the tyranny presented by the demands of 'progress'. From this perspective, counterurbanisation can be seen as part of a game, an 'experience' freed of any of the existential burdens which modernist logic might have placed upon it. More specifically, there are two ways in which we may seek to describe counterurbanisation as postmodern: first, as a relatively limited reaction against the tyranny of urban rationalism, an active form of the subversion suggested by Baudrillard's 'silent majority'; and, secondly, as a 'lifestyle' built on surface imagery.

It seems clear that counterurbanisation *does* represent a clear-cut rejection of urban life as it has typically been experienced by the migrants. As such, given the way in which the city and urbanism have tended to appear as leitmotifs of modernism (Gruffudd 1995: 51), we may thus wish to categorise the 'return to the rural' as postmodern. However, such an integration appears rather limited and unduly reliant upon empirical trends and spatial patterns at the expense of the fundamental cultural dimension of postmodernism. This observation is supported by the lack of the necessarily 'superficial' lifestyles among migrants. From personal experience of fieldwork and through a broader reading of the literature, the 'commitment' and seriousness of the migrants is immediately clear. In spite of the suggestions of some authors (e.g. Newby 1980; Thrift 1987, 1989, on the 'green welly–Laura Ashley–Range Rover' culture), for a majority of counterurbanisers it is not a case of rural-lifestyle-one-week-urban-lifestyle-the-next.

To illustrate this point in greater detail, the 'aesthetic' appeal of rural living is not suited to the 'anti-auratic' character of the postmodern cultural encounter (Lash and Urry 1987: 286). The rural landscapes of the migrants do not conform to the postmodern work of art (Featherstone 1988; Hassan 1985; Lash and Urry 1987: 286–7): they are unsuited to mechanical reproduction; they are certainly more contemplative than distracted in reception; and, while the rural is 'popular', this is a popularity rooted in 'high art' and Culture rather than in the mundane features of everyday life. Indeed, ruralism is much closer to the narrative, 'discursive' character of modernism than to the spectacular 'figural' character of postmodernism (Lash 1988: 313–14), and a non-postmodern interpretation of counterurbanisation follows from this.

Premodern or modern reaction

Counterurbanisation appears to be too 'deep' to conform to the demands of a postmodern action. Can we, therefore, regard it instead as a form of premodernist or even modernist reaction to the postmodern condition? As Harvey suggests, such a response might not be unexpected:

> But, as so often happens, the plunge into the maelstrom of ephemerality has provoked an explosion of opposed sentiments and tendencies.... Deeper questions of meaning and interpretation... arise. The greater the ephemerality the more pressing the need to discover or manufacture some kind of *eternal truth* that might lie therein.
>
> (Harvey 1989: 292; my emphasis)

Hence, counterurbanisation can be seen as part of a strategy to deny the postmodern complexity and to create and sustain 'eternal truths'. These may be rooted in either a premodernist or a modernist discourse. From such a perspective, counterurbanisation and its associated ruralism are part of an effort to explain the world through a series of simple slogans of the 'old truths' (O'Neill 1988).

Most support in the literature for such a reactionary interpretation of counterurbanisation regards it as drawing upon premodern sympathies, reflecting the long historical pedigree of ruralist attitudes within British society. The idea that the rural provides an 'escape' from an uncertain, multiracial and crime-ridden urban world into the 'timeless' countryside, with its social quietude, peace and beauty, is commonly expressed by counterurbanisers. Here, rurality's role as a 'refuge from modernity' (Short 1991: 34) in the guise of the city and its corresponding other 'old truths' of Enlightenment modernism is important. Associated with this consolatory role is often a nostalgia for the certainties of a rural childhood and other past rural experiences, since the countryside is associated not only with 'the past' but also with the personal past of childhood (Squire 1993; Williams 1973). Moreover, it is in childhood that many of our attitudes to the rural are formed (Harrison *et al.* 1986, 1987). Tuan (1980) and Olwig

(1982) argue that, with adulthood, an individual develops a sense of place which, in contrast to the unconsciousness of childhood rootedness, involves greater distancing between the individual and particular places. Thus, nostalgia for one's childhood can be seen as an attempt to develop a sense of place and ontological security through reflection upon a lost rootedness.

Furthermore, the generally 'conservative' character of the counterurbanisers, whether in terms of lifestyles or attitudes, is readily apparent from evidence presented by the bulk of migrants to rural areas. The 'ordinariness' of the 'new' rural inhabitants has also been emphasised (Bolton and Chalkley 1989). The migrants, even those from the service class, seem on the whole not to be the sort of people one would associate with any cultural 'vanguard', whether modernist or postmodernist.

Nevertheless, it seems too easy to dismiss counterurbanisation as a reactionary premodern nostalgic response. First, the migrants themselves would be the first to dispute this equation. Experience suggests that they do not see themselves, by and large, as living in the past. For example, many migrants are quite capable of seeing through the 'chocolate box' stereotype of the village presented by the 'rural idyll' (Halfacree 1995). Looking at the lifestyles of the counterurbanisers, one is often hard-pressed to see many elements of 'Old Fogeyism' or other attempts at living in the past. Moreover, awareness of the general 'green' debate and concern with matters such as traffic congestion and pollution illustrate how a number of very contemporary concerns may be woven into the counterurbanisers' ruralist discourse. Most crucially, however, we must acknowledge how seemingly reactionary images and ideas can to some extent escape their ideological shackles and resonate deeper and more general concerns. As Wright notes, albeit in a qualified manner, with respect both to the heritage industry and to the 'Deep England' of our national (sic) pastoral vision:

> Like the utopianism from which it draws, national heritage involves positive energies which certainly can't be written off as ideology. It engages hopes, dissatisfactions, feelings of tradition and freedom, but it tends to do so in a way that diverts these potentially disruptive energies into the separate and regulated spaces of stately display.... Deep England makes its appeal at the level of everyday life. In doing so it has the possibility of securing the self-understanding of the upper middle-class while at the same time speaking more inclusively in connection with all everyday life, where it finds a more general resonance.
>
> (Wright 1985: 78, 87)

Indeed, the very 'vagueness' (Wright 1985: 81) of Deep England enables it to speak for many often contrasting experiences, as can the imprecision of the community and structure of 'Wanet' in the Lake District allow some very diverse interpretations of that village by its inhabitants (Rapport

1993). Similarly, Harrison (1982, 1986) recognises a 'positive nostalgia', with an imagined past used actively as a utopian-like vision with which to critique the present.

Considering counterurbanisation to be a modernist reaction to the post-modern experience is harder to accept than a premodernist interpretation. On the one hand, with some notable exceptions (Gruffudd 1995; Matless 1994: 12–16), the rural is rarely the spatial location for 'progress' in either popular or academic discourse, as has already been noted. Furthermore, while one might not necessarily accept the idea that counterurbanisation is a premodern response, we are still left with the idea that the country-side is a well of premodernist sentiment and associations, making it hard for modernism to claim for its own, except perhaps in some post-Fordist sense. Thus, it is very hard to link counterurbanisation with some essentially *reactionary* modernism. Overall, therefore, counterurbanisation does not conform to the fundamentalism of Harvey's sloganeers but is better seen as representing an attempt at the contemporary (re)appropriation of the past in the cause of the present.

COUNTERURBANISATION IN A POSTMODERN CONTEXT: 2. A CRITICAL ALTERNATIVE?

Counterurbanisation and critical modernism

There is an alternative way of interpreting counterurbanisation which sees the migration trend as neither essentially postmodern nor reactionary but as a combination of a modernist search for order relativised by a postmodern acceptance of the boundedness of knowledgeability. This interpretation enables the suggestion to be made that counterurbanisation contains within it a potentially radical current. Such an understanding comes from Harvey's suggested strategy in dealing with the existential challenge provided by the postmodern turn, involving the attempt to produce a basis for limited action and understanding in everyday life. With this strategy, the individual attempts to re-place herself or himself in the world and regain a sense of perspective (Eyles and Evans 1987; Giddens 1990: 141), while accepting that the whole world can never be 'known' by any individual (also Castells 1983; Gregory 1989; Jameson 1988; Ryan 1988; Zukin 1988). In other words, this 'critical modernism' (or 'critical postmodernism') argues that we have to find our place without the help of the 'grand narratives' of the past or, at least, with self-reflexivity in the use of such narratives. Such an approach, after Bauman (1990), is also advocated by Smart's (1993) 'relational' postmodernism. Psychologically, we need to find our place in 'a world in which one is not necessarily a stranger' (Williams 1973: 298). Consequently, 'We need answers, even if they are justifiably recognised to have a relatively provisional, conditional, and even at times contentious status' (Smart 1993: 125).

Critical modernism can be seen as a form of dialectical synthesis between postmodernism and premodernism, refracted through the modernist concern for order. A search for ontological security through order can be linked as much to modernist sentiments (Matless 1990) as to premodernism (Wright 1985), for, as Smart (1993: 41) notes, 'The idea of order as a task, as a practice, as a condition to be reflected upon, preserved and nurtured is intrinsic to modernity.' Why counterurbanisation should provide a means of achieving such order reflects the strong link between ruralism and 'tradition', discussed above. As Friedman (1988: 452) argues, traditionalism provides the roots and values necessary in the absence of a modernist vision, where a loss of faith in the city as Jerusalem accompanies the loss of faith in collective emancipation (Short 1991: 90). Yet, crucially, as we saw in the previous section, traditionalism should be seen as being used creatively and not just providing a supposed escape. Instead of seeing the desire for openness, quietness, cleanliness, aesthetic quality and 'nature' – characteristics stressed by counterurbanisers (Halfacree 1994) – as reflecting a postmodern concern with style or a premodern concern with the past, we can see it as an attempt to create a 'distance' between the migrant and the rest of the world, as symbolised by the urban population, in order to overcome the 'de-differentiating' tendencies of postmodernism (Lash 1988). The rural physical landscape is more precisely defined and delineated – more ordered – than that of the city. Similarly, the desire for a residential escape, a slower pace of life, a sense of community, safety and familiarity, can be linked to a need to obtain a sense of belonging in the world as it is today, rather than to achieve a return to a past world or to play games of 'country living'. Rural society, particularly on account of its more intimate scale, is one in which migrants feel they can get a better overview; the rural social landscape, too, is more ordered.

The 'quality of life' so important in driving counterurbanisation thus can be seen as reflecting fundamental existential concerns. Just as Williams's (1973) cultural materialism demonstrated how the ideas of 'country' and 'city' in bourgeois British literature constantly changed to reflect the times in which they were produced, so too does contemporary ruralism reflect a way of interrogating the postmodern experience through regenerated and reconstituted 'traditional forms of life' (Smart 1993: 109; also Short 1991: 31). It is therefore unsurprising that counterurbanisation is undertaken by such a wide range of people and, in attempting to build up this sense of ontological security, we can appreciate the concern with rural and/or local ties and connections; the interest expressed by migrants in local events and traditions; and the tendency to migrate to rural areas prior to child rearing, given the positive associations between rurality and childhood.

Rejection of the 'grand narrative', finally, accords counterurbanisation's modernism its critical aspect, although this is harder to appreciate. None the less, we can start by pointing to the migrants' critical awareness of the 'rural idyll' of *Gemeinschaft* certainty and harmony, even when they reproduce this representation in their definitions of the rural (Halfacree 1995). Such an image may provide a 'blueprint' for the rural 'ideal' but it can be readily unmasked when discussed directly. Furthermore, there is awareness of perceived threats to rural life, whether expressed in terms of the physical encroachment of development or the social challenge posed by crime and television. Secondly, migrants are often aware that rural living is not 'for everybody', whether this selectivity is explained in terms of cultural choice or financial constraint, and thus does not provide any singular answer to questions of ontological insecurity. Thirdly, the very rejection of city living implies a recognition of the fallibility of the old certainties of progress. However, as is elaborated upon in the next section, the critical sense of perspective that counterurbanisation provides is certainly *not* always developed, thereby compromising counterurbanisation's critical edge.

The scope for a radicalism

Figure 4.1 demonstrates the precarious position of critical modernism, as it strives not to dissolve into either of the poles of postmodern relativism or premodern essentialism. The sympathies and potentials popularly expressed by counterurbanisation are equally torn. On the one hand, counterurbanisation and the attendant ruralism which it draws upon and (re)produces can slip towards the playful, style-obsessed 'trendiness' which plays up the act of migrating and living at the expense of the deeper sympathies and meanings which are trying to be expressed; aesthetic over ethic, Being over Becoming. On the other hand, the nostalgic character of rural living can assume precedence and the righteousness and morality of rural living can be trumpeted uncritically. Here, it is absolute ethics devoid of contextualisation which predominates. This section considers these twin dangers, starting first with the problem of premodernist laudatory perspectives on the rural.

The dangers of rural idolisation (idyll-isation?) and a failure to expose the limitations of the grand narrative are apparent in the debate over 'marginal' rural geographies which has informed much of this book. The key insight here is, of course, Chris Philo's (1992) 'Neglected rural geographies'. As a relative 'outsider' with respect to rural geography, Philo was able to note more clearly than many rural researchers the way in which the subject was biased towards the concerns of what he termed 'Mr Average' or: 'men in employment, earning enough to live, white and probably English, straight and somehow without sexuality, able in body and sound in mind, and devoid of any other quirks of (say) religious belief or

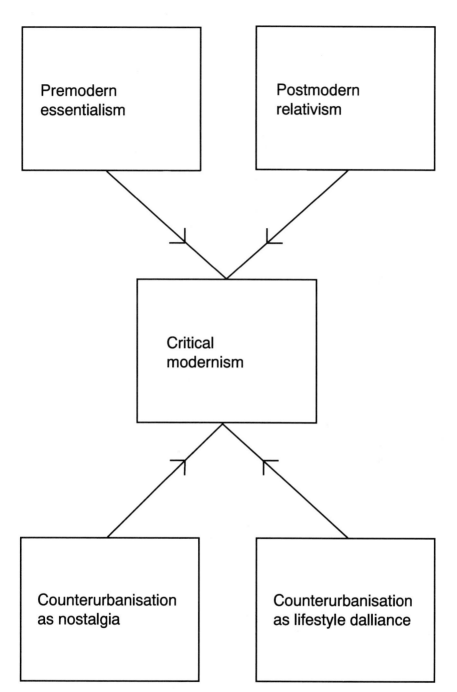

Figure 4.1 Locating counterurbanisation with respect to premodernism, postmodernism and critical modernism

political affiliation' (Philo 1992: 200). In other words, the rural population is made 'similar', if not 'the same', and any sense of 'otherness' is ignored or simply denied. Similarly, an uncritical acceptance of the rural idyll sees 'the rural' and 'deprivation' become contradictions in terms (Cloke 1994: 175).

Philo's concern with the need to study rural 'others' has been reflected elsewhere (for example, Cloke 1993, 1994; Roberts 1992), with his paper forming the subject of a debate between himself and Murdoch and Pratt (Murdoch and Pratt 1993, 1994; Philo 1993). The latter authors argued that it was not enough just to study rural 'others', as this could lead to a fragmented 'pick 'n' mix' approach which, *inter alia*, would tend to bypass the issues of power which consolidate, albeit temporarily and often not without struggle, the rural whole (Cloke 1994: 183). Instead, the issue of the 'production of the rural' must be considered directly.

From the perspective of counterurbanisation, we can appreciate immediately from this debate the dangers of unproblematically celebrating the rural as a potentially ideal residential environment. In short, while counterurbanisation might be the ideal strategy for Philo's Mr Average in his search for a basis from which to regain and retain a sense of ontological security, it is not an option for more marginalised groups. For them, the rural idyll is much less of an idyllic representation, prompting the challenge to that representation from groups such as Travellers (Halfacree 1996a) and many of the other sections of society discussed in this book. It is, of course, such groups which postmodernism has done so much to illuminate.

The dangers of the essentialisation of the countryside and of regarding counterurbanisation as an intrinsically progressive cultural response are mirrored by problems with adopting a more postmodern emphasis. While a postmodern emphasis might recognise the partiality of the 'counterurban solution' for society as a whole, over-relativism can lead to disillusion and crisis in its own right. Such an appreciation comes from Maffesoli's concept of 'neo-tribalism', which is central to his discussion of postmodern society. The 'neo-' here is crucial, as it allows a distinction to be made with traditional 'tribalism'. Maffesoli suggests that in the postmodern era new collectivities are emerging to replace class, defined in terms of a 'multitude of individual acts of *self-identification*' (Bauman 1992: 136). Individuals congregate in order to 'bathe in the affectual ambience' (Maffesoli 1991: 11) in their search for community and belonging. These postmodern neo-tribes differ from the *Gemeinschaft* communities of 'historical' tribes in that they are actively achieved by their individual members rather than being simply born into (Shields 1992: 14); they comprise a 'myriad of "unforced" communities' (Thrift 1992: 26; also Weeks 1993).

The survival of neo-tribes, given their 'worked at' origins, means that they have constantly to be actively monitored and reflected upon by their members. This perpetual need for self-monitoring means that neo-tribes

are unlikely to survive over long periods of time as the monitoring ultimately becomes too arduous. Self-monitoring also readily exposes the
impracticalities, contradictions and general insufficiency of neo-tribal
beliefs. None the less, it is suggested that neo-tribes can sustain themselves
over substantial periods of time if they are capable of creating their own
space (Halfacree 1997). Survival comes partly from Maffesoli's emphasis
on 'lived spaces' with their own 'emotional geographies', which suggests a
distinction between emotional geographies which remain in the 'imagination' and those which become materialised 'on the ground'. Drawing on
Lefebvre (1991), it is in the latter case that the neo-tribe will have the best
chance of sustaining itself. Lefebvre sums up the situation well:

> Any 'social existence' aspiring or claiming to be 'real', but failing to
> produce its own space, would be a strange entity, a very peculiar
> kind of abstraction unable to escape from the ideological or even the
> 'cultural' realm. It would fall to the level of folklore and sooner or
> later disappear altogether, thereby immediately losing its identity, its
> denomination and its feeble degree of reality.
>
> (Lefebvre 1991: 53)

To apply Maffesoli's ideas to counterurbanisation, it can be suggested
that this migration trend is part of the creation of a neo-tribal identity (or
possibly one from a set of neo-tribal identities) linked to ruralism and the
adopting of a 'rural lifestyle'. If this is so, then it becomes postmodern in
Maffesoli's terms. However, the ephemerality of neo-tribal existence suggests that counterurbanisers must produce a distinctive space if this neo-
tribal existence is to become consolidated and they are to engage with the
existential questions posed by the postmodern condition. Hence the
dangers of adopting a shallow, postmodern attitude to counterurbanisation
and rural living which fails to engage with the complexity of producing a
space. Instead, if the needs which are being expressed in the counterurbanisation decision are to be acted upon and met more completely, the
character and contours of the desired rural future need to be actively
thought about and planned. This will, of course, require reflection upon
the 'urban' existence which is being countered. It is here that the potentially
radical implications of counterurbanisation can emerge and be developed
if this engagement proceeds to become a critique of the experience of
contemporary capitalism.

CONCLUSION: ONE CRITICAL FUTURE?

As they used to say: only connect. Now the message is: only reconnect, but reconnect in a pluralistic world ... in Britain now, the chief
imperative is to find out how, why, with what and where people

belong: what their economic, social and cultural resources are and how they wield them. This is a necessary first step to establishing a new notion of collectivity, one which is not associated with uniformity and subordination but is 'active, demanding and creative' (Campbell 1988: 39). It remains to be seen whether this will occur.

(Thrift 1992: 27–8)

A novel, politically radical neo-tribal critique of contemporary society is currently taking place in Britain in the form of opposition to the government's road-building programme. Thus, for example, the campaigns against the M11 link in east London between 1993 and 1995 and the ongoing (1996) Newbury bypass protest are not just about roads and cannot be reduced to a NIMBY response to threats to property prices or personal quality of life. Instead, they have extended themselves to enter the political debate around community rights, land rights, economic policy, etc. For example, there is the newly launched campaign known as 'The Land Is Ours', which regards itself as a 'land rights movement for Britain' (The Land Is Ours 1995). The campaign aims to secure broader access to land in Britain, whether for homes, for employment or for popular access in general. George Monbiot, a key figure in the campaign, argues that the campaign's actions should not be seen as a protest *against* anything *per se* (Monbiot 1995) but as a reassertion of 'people's control over how they live' (*Guardian* 1995).

Whether the impulses behind counterurbanisation can be aligned with such radicalism remains to be seen. At first sight, it might appear most unlikely, given the gulf between 'respectable' in-migrant owner-occupiers and 'deviant' groups such as environmental protesters. However, observing the situation differently, there are similarities between many of the characteristics often associated with rurality and underpinning ruralism – such as community, quietude, beauty, harmony – and the world vision currently emerging in these environmental disputes. The major hurdle to overcome is the selective representation of the countryside – the rural idyll – within which the counterurbanisers' hopes and fears are typically enmeshed, and the lack of the necessary abstraction to be able to remove a recognition of the need for ontological security from this undemocratic spatial form. In short, counterurbanisation remains too close to the premodern pole in Figure 4.1 and has yet to become effectively synthesised with the postmodern pole.

Synthesising a postmodern sense of limits with the premodernism of counterurbanisation provides an opportunity for a sense of value to be added to the debate. Critical modernism enables us to explore the limits of the counterurban solution and to expose the source of the initial 'problem' (i.e. the postmodern condition). Moreover, it raises questions about the production of difference and thus returns debate to the 'moral' question of similarity (Harvey 1993). In short, some of the impulses behind both counterurbanisation and environmental protest come from a common source. There is similarity here, with many of the apparent differences between the two groups not reflecting significant degrees of 'otherness' but merely 'relational' otherness, linked with specific life-course experiences only (see Young 1993). To recognise any similarity provides an opening, although *just* an opening, for a future radical polity, for, as Harvey (1993: 114) argues, 'To discover the basis of similarity (rather than to presume sameness) is to uncover the basis for alliance formation between seemingly disparate groups.'

The detailed shape of the post-productivist countryside remains elusive. Perhaps the space opened up by the productivist–post-productivist shift will not be appropriated by the radical imagination. However, the post-Fordist vision of the countryside, although clearly still wedded to capitalism, does offer hope. For example, this vision attaches considerable significance to high-technology, information technology-driven economic operations, symbolised by telecottages and teleworking. Crucially, from a more social angle, these operations are almost always placed imaginatively in a village *community*, where a wide diversity of people engage with each other and with their work locally (for example, Toffler 1970; cf. Richardson *et al.* 1995). While this social vision is a powerful tool for marketing hype, it can perhaps also be seized upon by more radical elements and be used to argue for an inclusive rural future; a rural of Others rather than an exclusive rural of the Same.

Overall, then, if counterurbanisation can be harnessed in a critical modern way, rather than just as a 'lifestyle' or as nostalgic reaction, then, with some imagination, it can form a progressive part of the contested countryside cultures explored in this book. Furthermore, it can also have ramifications, both political and socio-geographical, beyond the rural sphere itself, engaging with the existential crisis presented to us all by society's postmodern turn; for, as Mormont (1987: 11) observes, 'The countryside is often considered to be a place where it is possible to put into practice another way of life ... or another model of social and economic organisation.'

NOTE

1 An earlier version of much of the remainder of this chapter, drawing more heavily upon empirical material and concentrating on 'ruralism' rather than counterurbanisation, is presented in Halfacree (1996b).

KEITH HALFACREE

REFERENCES

Baudrillard, J. (1983a) *Simulations*, New York: Semiotext(e).
—— (1983b) *In the Shadow of the Silent Majorities*, New York: Semiotext(e).
—— (1988) *America*, London: Verso.
Bauman, Z. (1992) *Intimations of Postmodernity*, London: Routledge.
Bell, D. (1973) *The Coming of Post-industrial Society*, London: Heinemann.
Bolton, N. and Chalkley, B. (1989) 'Counter-urbanisation – disposing of the myths', *Town and Country Planning* 58: 249–50.
Bourdieu, P. (1984) *Distinction*, London: Routledge and Kegan Paul.
Burgess, J. (1982) 'Selling places: environmental images for the executive', *Regional Studies* 16: 1–17.
Campbell, B. (1988) 'Clearing the decks', *Marxism Today* October: 34–9.
Castells, M. (1983) 'Crisis, planning, and the quality of life: managing the new historical relationships between space and society', *Society and Space* 1: 3–21.
Champion, A. (ed.) (1989) *Counterurbanisation*, London: Edward Arnold.
—— (1992) 'Urban and regional demographic trends in the developed world', *Urban Studies* 29: 461–82.
Cloke, P. (1993) 'On "problems and solutions": the reproduction of problems for rural communities in Britain during the 1980s', *Journal of Rural Studies* 9, 113–21.
—— (1994) '(En)culturing political economy: a life in the day of a "rural geographer"', in P. Cloke, M. Doel, D. Matless, M. Phillips and N. Thrift, *Writing the Rural*, London: Paul Chapman Publishing, pp. 149–90.
Cloke, P. and Goodwin, M. (1992) 'Conceptualizing countryside change: from post-Fordism to rural structured coherence', *Transactions of the Institute of British Geographers* NS 17: 321–36.
Connell, J. (1978) *The End of Tradition: Country Life in Central Surrey*, London: Routledge and Kegan Paul.
Cooke, P. (1988) 'Modernity, postmodernity and the city', *Theory, Culture and Society* 5: 475–92.
Dickens, P. (1989) 'Society, space and human nature', *Geoforum* 20: 219–34.
Eyles, J. and Evans, M. (1987) 'Popular consciousness, moral ideology, and locality', *Society and Space* 5: 39–71.
Featherstone, M. (1988) 'In pursuit of the postmodern: an introduction', *Theory, Culture and Society* 5: 195–215.
Fielding, A. (1982) 'Counterurbanisation in western Europe', *Progress in Planning* 17: 1–52.
Fothergill, S. and Gudgin, G. (1982) *Unequal Growth*, London: Heinemann.
Fothergill, S., Gudgin, G., Kitson, M. and Monk, S. (1985) 'Rural industrialization: trends and causes', in M. Healey and B. Ilbery (eds) *The Industrialization of the Countryside*, Norwich: Geo Books.
Friedman, J. (1988) 'Cultural logics of the global system: a sketch', *Theory, Culture and Society* 5: 447–60.
Gerth, H. and Wright Mills, C. (eds) (1970) *From Max Weber: Essays in Sociology*, London: Routledge and Kegan Paul.
Giddens, A. (1981) *A Contemporary Critique of Historical Materialism*, London: Macmillan.
—— (1990) *The Consequences of Modernity*, Cambridge: Polity.
—— (1991) *Modernity and Self-identity*, Cambridge: Polity.
Gracey, H. (1973) 'The home seekers: family residential mobility and urban growth', in P. Hall, H. Gracey, R. Drewett and R. Thomas (eds) *The Containment of Urban England*, London: Allen and Unwin.

90

Gregory, D. (1989) 'Presences and absences: time-space relations and structuration theory', in D. Held and J. Thompson (eds) *Social Theory of Modern Society*, Cambridge: Cambridge University Press, pp. 185–214.

Gruffudd, P. (1995) ' "A crusade against consumption": environment, health and social reform in Wales, 1900–1939', *Journal of Historical Geography* 21: 39–54.

Guardian (1995) 'Protest groups unite to reclaim their lost land', 24 April.

Halfacree, K. (1994) 'The importance of "the rural" in the constitution of counter-urbanization: evidence from England in the 1980s', *Sociologia Ruralis* 34: 164–89.

—— (1995) 'Talking about rurality: social representations of the rural as expressed by residents of six English parishes', *Journal of Rural Studies* 11: 1–20.

—— (1996a) 'Out of place in the country: travellers and the "rural idyll" ', *Antipode* 28: 42–72.

—— (1996b) 'Ruralism and the postmodern experience: some evidence from England in the late 1980s', in M. Gentileschi and R. King (eds) *Questioni di popolazione in Europa: Una Prospecttiva Geografica*, Bologna: Pátron Editore, pp. 163–78.

—— (forthcoming 1997) 'Neo-tribes, migration and the post-productivist country-side', in P. Boyle and K. Halfacree (eds) *Migration into Rural Areas: Theories and Issues*, Chichester: Wiley.

Harrison, C., Limb, M. and Burgess, J. (1986) 'Recreation 2000: views of the country from the city', *Landscape Research* 11: 19–24.

Harrison, C., Burgess, J. and Limb, M. (1987) 'Popular values for the countryside', in B. Brown (ed.) *Leisure and the Environment*, Leisure Studies Association, Conference Papers no. 31, pp. 43–57.

Harrison, F. (1982) *Strange Land. The Countryside: Myth and Reality*, London: Sidgwick and Jackson.

—— (1986) *The Living Landscape*, London: Pluto.

Harvey, D. (1989) *The Condition of Postmodernity*, Oxford: Blackwell.

—— (1993) 'Class relations, social justice and the politics of difference', in J. Squires (ed.) *Principled Positions*, London: Lawrence and Wishart, pp. 85–120.

Hassan, I. (1985) 'The culture of postmodernism', *Theory, Culture and Society* 2: 119–32.

Jameson, F. (1984) 'Postmodernism, or the cultural logic of late capitalism', *New Left Review* 146: 53–93.

—— (1988) 'Cognitive mapping', in C. Nelson and L. Grossberg (eds) *Marxism and the Interpretation of Culture*, London: Macmillan, pp. 347–60.

King, A. (1989) 'The countryside: our hopes and fears', *Daily Telegraph* 14 August.

Kroker, A. (1985) 'Baudrillard's Marx', *Theory, Culture and Society* 2: 69–84

Lash, S. (1988) 'Discourse or figure? Postmodernism as a "regime of signification" ', *Theory, Culture and Society* 5: 311–36.

Lash, S. and Urry, J. (1987) *The End of Organized Capitalism*, Cambridge: Polity.

Lefebvre, H. (1991) *The Production of Space*, Oxford: Blackwell.

Levi, L. and Andersson, L. (1975) 'Population, environment and the quality of life', *Ekistics* 236: 12–19.

Lowe, P. and Goyder, J. (1983) *Environmental Groups in Politics*, London: Allen and Unwin.

Lyotard, J.-F. (1984) *The Postmodern Condition*, Manchester: Manchester University Press.

Maffesoli, M. (1991) 'The ethic of aesthetics', *Theory, Culture and Society* 8: 7–20.

Marsden, T., Murdoch, J., Lowe, P., Munton, R. and Flynn, A. (1993) *Constructing the Countryside*, London: UCL Press.

Matless, D. (1990) 'Definitions of England, 1928–89: preservation, modernism and the nature of the nation', *Built Environment* 16: 179–91.

—— (1994) 'Doing the English village, 1945–90: an essay in imaginative geography', in P. Cloke, M. Doel, D. Matless, M. Phillips and N. Thrift, *Writing the Rural*, London: Paul Chapman Publishing, pp. 7–88.

Meyrowitz, J. (1985) *No Sense of Place: The Impact of Electronic Media on Social Behavior*, New York: Oxford University Press.

Monbiot, G. (1995) 'Whose land?', *The Guardian* 22 February.

Mormont, M. (1987) 'Rural nature and urban natures', *Sociologia Ruralis* 27: 3–20.

Moseley, M. (1984) 'The revival of rural areas in advanced economies: a review of some causes and consequences', *Geoforum* 15: 447–56.

Mulgan, G. (1989) 'That certain feeling', *Marxism Today* September: 26–33.

Murdoch, J. and Marsden, T. (1994) *Reconstituting Rurality*, London: UCL Press.

Murdoch, J. and Pratt, A. (1993) 'Rural studies: modernisn, postmodernism and the "post-rural"', *Journal of Rural Studies* 9: 411–27.

—— (1994) 'Rural studies of power and the power of rural studies: a reply to Philo', *Journal of Rural Studies* 10: 83–7.

Newby, H. (1980) 'A one-eyed look at the country', *New Society* 53: 324–5.

—— (1987) *Country Life*, London: Weidenfeld and Nicolson.

Olwig, K. (1982) 'Education and the sense of place', in D. Cosgrove (ed.) *Geography and the Humanities*, Loughborough University of Technology, Occasional Paper 5, pp. 38–53.

O'Neill, J. (1988) 'Religion and postmodernism: the Durkheimian bond in Bell and Jameson', *Theory, Culture and Society* 5: 493–508.

Pfeil, F. (1988) 'Postmodernism as a "structure of feeling"', in C. Nelson and L. Grossberg (eds) *Marxism and the Interpretation of Culture*, London: Macmillan, pp. 381–403.

Philo, C. (1992) 'Neglected rural geographies: a review', *Journal of Rural Studies* 8: 193–207.

—— (1993) 'Postmodern rural geography? A reply to Murdoch and Pratt', *Journal of Rural Studies* 9: 429–36.

Pocock, D. and Hudson, R. (1978) *Images of the Urban Environment*, London: Macmillan.

Pollert, A. (1988) 'Dismantling flexibility', *Capital and Class* 34: 42–75.

Potts, A. (1989) ' "Constable country" between the wars', in R. Samuel (ed.) *Patriotism*. Vol. 3. *National Fictions*, London: Routledge, pp. 160–86.

Radford, E. (1970) *The New Villagers*, Birmingham: Frank Cass.

Rapport, N. (1993) *Diverse World-views in an English Village*, Edinburgh: Edinburgh University Press.

Richardson, R., Gillespie, A. and Cornford, J. (1995) 'Low marks for rural home work', *Town and Country Planning* 64: 82–4.

Roberts, L. (1992) 'A rough guide to rurality', *Talking Point* 137.

Ryan, M. (1988) 'Postmodern politics', *Theory, Culture and Society* 5: 559–76.

Sack, R. (1988) 'The consumer's world: place as context', *Annals of the Association of American Geographers* 78: 642–64.

Sahlins, M. (1976) *Culture and Practical Reason*, London: University of Chicago Press.

Samuel, R. (1989) 'Introduction: exciting to be English', in R. Samuel (ed.) *Patriotism*. Vol. 3. *National Fictions*, London: Routledge, pp. xviii–lxvii.

Sayer, A. (1989) 'Post-Fordism in question', *International Journal of Urban and Regional Research* 13: 666–95.

Sherwood, K. (1984) 'Population turnover, migration and social change in the rural environment: a geographical study of south Northants', unpublished PhD thesis, University of Leicester.

Shields, R. (1992) 'Spaces for the subject of consumption', in R. Shields (ed.) *Lifestyle Shopping: The Subject of Consumption*, London: Routledge, pp. 1–20.

Short, J. (1991) *Imagined Country*, London: Routledge.

Short, J., Witt, S. and Fleming, S. (1987) 'Conflict and compromise in the built environment: housebuilding in central Berkshire', *Transactions of the Institute of British Geographers* NS 12: 29–42.

Simmel, G. (1971) 'The metropolis and mental life', in D. Levine (ed.) *On Individuality and Social Form*, Chicago: University of Chicago Press, pp. 324–39.

—— (1978) *The Philosophy of Money*, London: Routledge and Kegan Paul.

Smart, B. (1993) *Postmodernity*, London: Routledge.

Smith, N. (1984) *Uneven Development*, Oxford: Blackwell.

Soper, K. (1981) *On Human Needs*, Brighton: Harvester Press.

—— (1993) 'Postmodernism, subjectivity and the question of value', in J. Squires (ed.) *Principled Positions*, London: Lawrence and Wishart, pp. 17–30.

Squire, S. (1993) 'Valuing countryside: reflections on Beatrix Potter tourism', *Area* 25: 5–10.

The Land Is Ours (1995) *Newsletter* 1.

Thrift, N. (1987) 'Introduction: the geography of late twentieth-century class formation', in N. Thrift and P. Williams (eds) *Class and Space*, London: Routledge and Kegan Paul, pp. 207–53.

—— (1989) 'Images of social change', in C. Hamnett, L. McDowell and P. Sarre (eds) *The Changing Social Structure*, London: Sage/Open University Press, pp. 12–42.

—— (1992) 'Light out of darkness? Critical social theory in 1980s Britain', in P. Cloke (ed.) *Policy and Change in Thatcher's Britain*, Oxford: Pergamon Press, pp. 1–32.

—— (1994) 'Inhuman geographies: landscapes of speed, light and power', in P. Cloke, M. Doel, D. Matless, M. Phillips and N. Thrift, *Writing the Rural*, London: Paul Chapman Publishing, pp. 191–248.

Toffler, A. (1970) *Future Shock*, New York: Bantam.

Townsend, A. (1993) 'The urban–rural cycle in the Thatcher growth years', *Transactions of the Institute of British Geographers* NS 18: 207–21.

Tuan, Y.-F. (1980) 'Rootedness versus sense of place', *Landscape* 24: 3–8.

Ward, N. (1993) 'The agricultural treadmill and the rural environment in the post-productivist era', *Sociologia Ruralis* 33: 348–64.

Weeks, J. (1993) 'Rediscovering values', in J. Squires (ed.) *Principled Positions*, London: Lawrence and Wishart, pp. 189–211.

Williams, J. (1981) 'The nonchanging determinants of nonmetropolitan migration', *Rural Sociology*, 46: 183–202.

Williams, R. (1973) *The Country and the City*, London: Hogarth Press.

Williamson, J. (1978) *Decoding Advertisements*, London: Marion Boyars.

Wright, P. (1985) *On Living in an Old Country*, London: Verso.

Young, I. (1993) 'Together in difference: transforming the logic of group conflict', in J. Squires (ed.) *Principled Positions*, London: Lawrence and Wishart, pp. 121–50.

Zukin, S. (1988) 'The postmodern debate over urban form', *Theory, Culture and Society* 5: 431–46.

5

ANTI-IDYLL
Rural horror
David Bell

Ever since there has been a clear distinction between the country and the city, the rural has occupied an ambivalent position in the popular imagination. On the one hand it is cherished as an innocent idyll of bucolic tranquillity and communion with nature – a place to retreat from the ever-quickening pace of urban living and to join in with 'authentic', rustic community life. From industrial barons buying up country estates and grand mansions in the Victorian countryside to current population movements characterised (not without contestation) as counterurbanisation, the appeal of the countryside remains strongly magnetic for those with sufficient mobility; and, of course, revolutions in temporary mobility (especially the automobile) have given almost everyone at least limited access to the great outdoors (Urry 1995). As a resource in popular culture texts, too, the 'armchair countryside' (Bunce 1994) offers countless ways of escape for urban dreamers to consume.

Against these idyllic representations of rurality, however, can be set other visions of the rural. Newcomers to the rural have always found aspects of country life less than Edenic, for example – from *arriviste* landowners confronted by nature red in tooth and claw (Rubenstein 1981) to 1980s commuter-villagers thrown into panic by tabloid folk-devil stories about 'rural rowdies', city folk have often found the countryside positively dystopian. In the USA, narratives of rural regions like the 'Deep South', such as those contained in the gothic writings of Flannery O'Connor, have long explored the badlands of the rural; its sick, sordid, malevolent, *nasty* underbelly.

In this chapter I want to focus on particular American filmic representations of this underbelly, which I group together here under the heading 'Rural horror'. By concentrating on issues of the representation of the rural, I hope to expand debates about the countryside as an 'imaginative resource' (see Bell and Valentine 1995). By thinking about rurality as an imagined cultural geography, we can examine issues of, among other things, the valencies given to the rural and the urban as popular cultural tropes –

94

and begin in to see how the interplay of lived cultures and cultural 'texts' works and reworks our understandings of what those tropes mean.

Since I shall be relying in this chapter primarily on North American cinematic portrayals of the countryside, it might be worth making a few preliminary comments about the national construction of the rural idyll in the UK and the USA. As Michael Bunce (1994) outlines, these 'idealised countrysides' have quite different origins, and are therefore represented in quite different ways – though there are notable commonalities, not least in the overbearing nostalgia attached to past rural cultures. In Britain, the rural idyll is a settled landscape mapping out a social order across a picturesque terrain – especially its construction as 'village England' (Cloke and Milbourne 1992). As Bunce (1994: 35) notes, 'the word countryside has not carried with it the same emotional connotations in North America as it has in Britain'. In the USA, the small town and the farm are the valued landscape forms of the rural – although the wilderness has an important role to play, too, with its associations of frontier life and the 'Old West'.

These constructions and representations of the idyllic rural have a long tradition both in Britain and in North America – from Henry David Thoreau and Walt Whitman to contemporary television shows like *Northern Exposure* and films such as *A River Runs Through It* or *Grand Canyon*, where all the problems of contemporary urban dis-ease are cured by an awe-filled trip to the wilderness landscape of the Grand Canyon (for an interesting discussion of *Grand Canyon*, see Giroux 1994); and from Thomas Hardy and Edward Carpenter to *Heartbeat* and *The Archers*. As a restorative to urban anomie, the rural has also gained prominence within, for example, the ecofeminist and mythopoetic men's movements. Against this have developed both criticisms of the exclusivity of the rural idyll (e.g. Halfacree 1996; Philo 1992), and a series of anti-idyllic visions of the rural, perhaps most explicitly played out in horror genres. And in response to the twee pastoralisms detailed by Bunce (1994) in his readings of depictions of the rural in popular culture, I want to offer not the 'armchair countryside' but the *behind the sofa countryside*; a place far, far from idyllic. Within the context of current academic debates on the social meanings and codes of horror film genres – and on ways of 'writing' and 'reading' the rural as a cultural 'text' – my discussion centres on articulations of the rural (and the urban) through the plotlines and characters of a variety of horror movies in which the 'victim' is coded as urban and both the setting and the 'monster' as rural.[1]

Horror films have been subject to considerable recent academic scrutiny. Interest in them is usually concerned either with broadly semiotic readings of the 'meanings' of horror (as encoded by its makers), or with the interpretation and uses audiences make of the horror 'text' (e.g. Crane 1994; Grant 1984). A significant flurry of intellectual activity has taken as its focus articulations of gender, sex and sexuality (e.g. Creed 1993; Clover 1992) and discourses of the body (e.g. Creed 1995; Finlay 1995) in horror. My concern here is with the uses of a particular binary – urban–rural – and in the codings around that which pitch 'innocent' urban newcomers against the visceral evils of the rural; films which Carol Clover (1992) terms 'urbanoid'.

The countryside has always loomed large in the horror genres as the perfect backdrop for terror. A vast repertoire of eerie or evil locations for the unfolding nastiness includes deserted mansions (in countless vampire films); the cottage in the country which urban travellers (lost, or let down by their car) take futile refuge in (memorably used in *Evil Dead*); whole villages in which evil runs amok, and into which witless urbanites stumble or are ensnared (see the brief discussion of *Two Thousand Maniacs* below); and endless permutations of city folk transplanted to the country on camping trips, or out hiking, touring or exploring, or at summer camp (as in *Friday the 13th*, also discussed below); or, like the soldiers in *Southern Comfort*, taken in some way or another out of their familiar environment and into the misty miseries of the swamplands (or forests, or whatever) . . . the list goes on. Of course, one of the main reasons for deploying the rural as a location is that it offers isolation and an alien environment; but this is also routinely achieved in other settings: at sea in *Dead Calm* or below it in *The Abyss*; in space (used very effectively in the *Alien* trilogy), a post-apocalyptic landscape (*Mad Max*) or a foreign land (travellers to the Haiti beware voodoo, as in *The Deep*). And even the most sprawling megalopolis can be the most threatening of places to get lost in or be pursued through (see, for example, *Candyman*, or the terror-filled suburbs in the *Nightmare on Elm Street* series). However, the films I will be discussing in detail in this chapter do not purely use the rural as the backdrop for their horror stories; they also all use country folk in the role of monster or murderer (usually a hybrid of both). Whereas in the classical Hollywood horror film (such as Universal Studios' Frankenstein movies), the monster invades the village and it is up to the villagers to pull together and defeat it (often drawing on folkloric local knowledge, community cohesion and resource-pooling; see Crane 1994), in the films featured below the villagers *are* the monsters, and they often cannot be defeated.

The locations in which the movies are based have spawned a peculiar species, usually referred to in US horror films as hillbillies, rednecks or mountain men. Trading on assorted cultural myths – of inbreeding, insularity, backwardness, sexual perversion (especially incest and bestiality) – these rural 'white trash' are familiar popular culture icons, celebrated by rock and pop bands such as Rednex and Southern Culture on the Skids and depicted in more benign ways in television shows and movies like *The Dukes of Hazzard* and *The Beverly Hillbillies*. The comic stereotype of the 'village idiot' prevalent in British humour is the UK equivalent of what

Joshua Glenn (1995: 46) calls these 'imbecilic unwashed mountain geeks'.[2] Their deployment in horror films, I argue, is as symptoms of social, cultural and economic processes having profound impact upon rural regions in the USA – of cultures, as the bandname suggests, on the skids.

HILLBILLY HORROR AND THE RURAL SLASHER

My particular interest here is in how particular country folk are depicted in horror movies, and how the clash of cultures between their brand of the rural and the urbanites whom they come into (usually deadly) contact with is written as 'hillbilly horror'. The discussion which follows concentrates largely – although not exclusively – on so-called 'slasher' movies (also called stalk 'n' slash, splatter, shocker or meat movies). The genealogy of the genre is now well rehearsed, its common plotlines described and analysed in a number of texts (see, for example, Clover 1989, 1992). Initially viewed as quintessentially 'low' cinema – as being 'drenched in taboo and encroaching vigorously on the pornographic' (Clover 1989: 91) – the slasher film has come to be regarded by some as an important commentary on (American) culture and society, often containing progressive, transgressive and radical critique (e.g. Modleski 1986; Sconce 1995). I want briefly to talk here about two films which embody common tropes of the 'rural slasher' film – *Two Thousand Maniacs* and *Friday the 13th* – and then look in much more detail at perhaps the archetypal slasher, *Texas Chainsaw Massacre*, directed by Tobe Hooper. The final film I will look at is from a very different orientation, combining rite-of-passage, man-against-nature, thriller and actioner status: John Boorman's *Deliverance*.

All these films, though with some variation, deploy particular country characters in writing their rural horrors. In American slasher films these characters – hillbillies, mountain men, Cajuns, rednecks – prey on assorted city-dwellers who enter their territory. In films such as *The Hills Have Eyes* and endless copies, sequels and variations, hapless city folk meet bloody ends at the hands of individual, gangs and (most often) families of killer hillbillies. As a cinematic type, the murderous yokel – even if transplanted to a more urban location, as in the case of Maynard and Zed, the butch pawnshop sodomites in *Pulp Fiction* – articulates perfectly one notable contested countryside culture, wherein the country wreaks its revenge on the city. (In contrast, Carol Clover (1992) discusses the rape-revenge nasty *I Spit on Your Grave* as the city getting even with the country.)

Academic work, mainly from anthropological and ethno-cultural studies, confirms the continued existence of traditional rural communities across parts of the USA which have retained a distinct identity through isolation, even in the midst of industrialisation. In the 1940s, the Kinsey Report on American sexual behaviours captured something of their lifestyles, especially in the phrase used of men involved in jobs like lumberjacking or cattle-running, for whom same-sex activity was surprisingly common: 'these men have faced the rigors of nature in the wild. They live on realities and a minimum of theory' (Kinsey *et al.* 1948: 457). In accordance with the stereotype, the report also highlighted the occurrence of bestiality and incest in rural America. Other ethnographic studies of particular communities give considerable insight into their ways. For example, residents of the New Jersey Pine Barrens (Pineys) continue to eke out a self-sufficient livelihood and maintain a culture based on city/country dichotomies played out in, among other things, culinary practices (Gillespie 1984). Similarly, the Cajuns of south Louisiana celebrate cultural distinctiveness, and even radicalise it as 'Proud Coonasses' raised on crawfish (Gutierrez 1984). The point where 'local patriotism' merges into xenophobia is the point where the slasher movie sits – where strangers are not merely avoided, but erased.

The strangers erased are always incomers from the city, either in search of the pleasures and adventures offered by the wilds, or simply lost, misdirected or redirected on their travels. In *Two Thousand Maniacs*, a north–south dichotomy is used to amplify the urban–rural divide. Highway travellers with Northern licence plates are rerouted by two villagers (Sconce (1995) calls them 'psychotic hayseeds'), and find themselves in the heart of Pleasant Valley, at the start of its centenary festivities. Their involvement in the celebrations soon becomes clear: one is barbecued, another drawn and quartered, the third rolled downhill in a barrel spiked with nails, and the last crushed beneath a huge boulder (see Finlay 1995 on the 'destruction of the body' in *Two Thousand Maniacs*). Here is local pride reaching its sadistic limit in very familiar slasher motifs. In her classification of slasher narratives, Carol Clover (1989) calls the site of the action 'Terrible Place'. The ironically, comically named Pleasant Valley is just such a place.[3]

A second common meat movie plotline lets the city victims simply stumble, quite by chance, into their unfolding nightmare. The highway is not only a place to be malevolently diverted off; it is also a place to get lost on, to break down on (beware of flagging down passing motorists in beat-up old station wagons), or to turn off on some adventure or errand (in *Texas Chainsaw Massacre*, the travellers are in search of their grandfather's grave). Others take off for outdoor pursuits (but not of the kind they end up partaking of); camping is a hazardous pastime on a slasher set. Typically, a group of happy-go-lucky city kids pitch tents in the woods, out of contact with the wider world. In a kind of anti-frontier myth, they find themselves inept at civilising nature; it becomes uncontrollable, alien, terrifying and ultimately murderous (horror films featuring animals or animal–human hybrids take this theme further; see Creed 1995).

The action in *Friday the 13th*, made in 1980, takes place in Camp Crystal Lake, an all-American children's summer camp. The victims are the camp counsellors, who seemingly function adequately together as a transplanted community, sharing chores and tasks, and apparently bonding with each other in Edenic surroundings. When the horrors of the film begin, however, and the killer, Jason, begins his work, the community rapidly disintegrates:

> At the first sign of danger, the crew is fractured into a bunch of helpless and weak individuals who have no hope of survival.... The film takes the nascent community ... and crushes it. Victims, gifted or without special talents, number so many bags of muscle and viscera for Jason to slash to pieces.
>
> (Crane 1994: 146)

The bulk of the film is then concerned with dispatching the roster of counsellors in various locations around the camp until the last one, Alice – the 'Final Girl' in Clover's (1989) taxonomy of slasher plots – turns the tables and slaughters Jason, the masked killer, who, we are surprised to discover, is in fact a woman, Mrs Voorhees. The killings are revenge for the accidental drowning of Mrs Voorhees' son – Jason – thanks to the negligence of former camp counsellors.

What is more interesting to us here, on top of the critique of young urban adults attempting to build up a sense of community, is that in most of the movie's many sequels, Jason himself becomes the killer: he had not drowned after all, but gone native, and ended up living in a hut in the woods surrounded by the kinds of perverse paraphernalia we have come to expect from a rural slasher (including evidence of fetishistic cannibalism, and a shrine built around the severed head of his mother, evidently retrieved from her death scene in Part One).[4]

So, in *Two Thousand Maniacs* and *Friday the 13th* we have two archetypal rural slasher tropes. In both, as in *Texas Chainsaw Massacre* and *Deliverance*, the horror is already there, waiting for the city folk to arrive. The Pleasant Valley villagers rerouting highway travellers and Jason/Mrs Voorhees are both slashers waiting to happen; like anthropologists stumbling across a lost tribe, the urbanites meet an alien culture where the norms of their own society count for nothing. Their innocent picture of the countryside – as a place passed through or to holiday in – is forever bloodied. Bloodier than most is the fate that awaits the hapless city youths who fall victim to the horrors of a film described variously as 'the *Gone with the Wind* of meat movies' and as 'a vile little piece of sick crap ... as extreme and hideous as a complete lack of imagination can possibly make it' (quoted in Clover 1989: 91–2), Tobe Hooper's *Texas Chainsaw Massacre*.

TEXAS CHAINSAW MASSACRE

Hooper's movie, made in 1974, is a landmark in the slasher genre, virtually defining its subsequent forms (many following slashers make more or less reverential reference to the film). Its plot has a group of innocent, young city kids – Sally, Jerry, Kirk, Pam and Franklyn – drive through east Texas for unspecified reasons (off on holiday perhaps, or just enjoying being teenagers on the open road). A radio report detailing grotesque goings-on in a local cemetery leads them to detour, in order to check on Franklyn and Sally's grandfather's grave (tying their lineage back to the country they have now left for city life). The landscape they pass through is cattle country, in which the mechanisation of slaughter has profoundly affected local job prospects (leading Robin Wood (1984) to read *Texas Chainsaw Massacre* as a critique of the effects of capitalist 'progress').

Before long they stop to pick up a hitchhiking local yokel. But when he mutilates both himself and Franklyn, then starts a fire in the back of the van, they hastily drop him off. Rather than turning for home they drive on, in good ol' slasher tradition, and end up at Franklyn and Sally's grandparents' deserted and ruined mansion (clearly echoing Bates' Motel from *Psycho*). As the kids explore the neighbourhood and Sally reminisces on her childhood, the landscape slowly becomes more and more . . . weird. Next door is a farmhouse, and Kirk goes in, in search of gasoline (running out of fuel in Terrible Place is another slasher staple). Inside he is savagely slaughtered by a freak wearing a slaughterhouse apron and a bizarre face mask (he is referred to in the credits as Leatherface). Leatherface's male family (his brother is the hitcher, and he is joined by his father and his barely alive grandfather; later we also 'meet' the mummified corpse of his grandmother, who still sits in her chair) represent 'three generations of slaughterhouse workers, once proud of their craft but now displaced by machines', who have resorted to 'killing and cannibalism as a way of life' (Clover 1989: 96). Together they massacre the other kids, leaving Sally as sole survivor. She is tortured, but manages to escape after a terrifying series of chases and near misses. A passing trucker (surprisingly, not a slaughterhouse family member) picks her up, and the film closes with Leatherface on the highway, maniacally waving his chainsaw aloft.

In a detailed reading of *Texas Chainsaw Massacre* as an apocalyptic allegory, Christopher Sharrett echoes Wood's comments about country people marginalised by the onward march of capitalism; in his view, the film shows 'the violent disruption of the security and stability of rural and suburban life' (Sharrett 1984: 263). As he concludes,

> Hooper's apocalyptic landscape is . . . a deserted wasteland of dissolution where once vibrant myth [of frontier] is dessicated. The ideas and iconography of Cooper, Bret Harte, and Francis Parkman are now transmogrified into yards of dying cattle, abandoned gasoline stations, defiled graveyards, crumbling mansions, and a ramshackle farmhouse of psychotic killers. *The Texas Chainsaw Massacre* [is] . . .

recognizable as a statement about the end-time of American experience.

(Sharrett 1984: 272)

The frontier-myth inversions of the film are also a hallmark of the rural slasher, but the themes of cannibalism and the fetishisation of human remains (furniture made from body parts, Leatherface's human-skin mask) also point to Nazi death camps. In Sharrett's words, 'the civilizing spirit has run its course; its energies are depleted, its myths not only dead but inverted and forced to show the consequences of their motivating force' (271). In fact, the film emphasises the very thin line between the romanticised frontiersmen and the sadistic cannibals.

The 'true crimes' upon which *Texas Chainsaw Massacre* is broadly based (and upon which countless other directors, including Hitchcock, have drawn) were the Ed Gein murders in 1950s Wisconsin. Gein, a handyman-turned-mass murderer and cannibal, was read by commentators as the by-product of an environment gone off the rails. A subsequent photo essay, *Wisconsin Death Trip* (Lesy 1973), evocatively set idyllic pastoral images of past Wisconsin residents against a list of bizarre crimes enacted in the state, and, by extension, across rural America. The inference – and the link to *Texas Chainsaw Massacre* – is clear: the lineage of rural horror is a long one, its current incarnation (whether Ed Gein or Leatherface) only its latest manifestation and a reaction to the society of the day. The themes of repetition and routinisation, also played out in *Texas Chainsaw Massacre*, further emphasise the everydayness of the atrocities (something also signalled in work on Nazi death camps as a phenomenon of modernity). Thus, the long-term shifts in American culture, society and the US economy have together conspired to spawn many monsters, of which the rural 'underwolf' is a long-running archetype.

Before we move on to look at *Deliverance* and finally draw together some conclusions, it should be mentioned here that Hooper's film was followed more than a decade later by a sequel. In *Texas Chainsaw Massacre II*, made in 1986, the family (now punningly named the Sawyers) have started an ominous (and successful) sausage business. The story centres on a local disc jockey, a woman named Stretch, whom the Sawyers suspect of knowing their secret recipe. Leatherface and his brother Choptop (the hitcher from Hooper's film) fail to kill her (largely through technical problems – Leatherface's chainsaw packs up), and are themselves ultimately, if rather predictably, dispatched by their own chainsaw (Clover (1989) makes an interesting reading of *Texas Chainsaw Massacre II*, especially its Final Girl sequence and Leatherface's loss of bloodlust, which she equates with his sexual awakening and attachment to Stretch – sex and violence are not mixed in the *Texas Chainsaw Massacre* films, unlike in many other sexploitation slashers).

The important themes of *Texas Chainsaw Massacre* in this context – inverted frontier myths, notions of a hopelessly corrupt and festering society, questions of civilisation and its discontents – are also played out, albeit in a very different genre, in the last film discussed here, *Deliverance*, which is notable for its line-up of stars (slashers usually use unknown actors – although many subsequently rise to celebrity status). Burt Reynolds, Jon Voigt, Ned Beatty and Ronny Cox play Lewis, Ed, Bobby and Drew, four men embarking on a nightmare journey in Georgia's backwaters.

DELIVERANCE

John Boorman's *Deliverance*, from 1972, comes from a very different tradition from the meat movie. It is an epic tale of four middle-class urban American men, led by Lewis, who embark on a canoe trip from Oree to Aintry, down a Georgia river soon to be flooded to create a hydroelectric dam system. Lewis is a survivalist, an Iron John character who wants to ride the rapids 'because they're there' (but soon won't be). He takes along Ed, Bobby and Drew, who are all clearly out of their favoured, cosy habitat (Lewis eggs them on to enjoy the trip, but they often seem reluctant; Bobby is described as a 'soft city, country club' type). Early on in the film, they try to arrange for some locals to drive their cars on to a prearranged pick-up point downriver. At first the men make jokes about the raw hillbilly community they have entered; but the more they see, the more horrified and (Lewis excepted) the more afraid they become. (In an intensely derivative *Deliverance* copy called *Hunter's Blood*, set in Arkansas, the city boys take photographs of the locals, one of them calling gleefully, 'It's like something outta *National Geographic*!' – and their extreme tourist gaze soon becomes their downfall, as they are chased out of town and into unknown territory by a wagon full of rednecks.)

The Appalachian landscape of *Deliverance* is not unlike that of *Texas Chainsaw Massacre*, only without the cannibal-fetish objects: a grim, beaten terrain of decaying car carcasses, ramshackle shacks and discarded farming equipment (allusions to poultry processing mirror the cattle-slaughter landscape of Texas). But, as with the kids' encounter with the slaughtering Sawyers, it is the *people* who most horrify the city folk. Peeping into a rundown hut, they see a wrinkled grandmother watching over a deformed child. The local men they meet are all stereotypical mountain hicks – toothless, imbecilic, unwashed (in the excellent *Hillbillyland* (1995), Williamson discusses Boorman and Dickey's notes on selecting the extras to play these bit parts, and the subsequent outcry from the locals of Rabun County, Georgia, on seeing themselves on screen).

In a key scene before the men set off, Drew plays a guitar and banjo duet with a local boy, Lonnie. Lonnie is a clear product of a society at the end of its existence, in-bred, retarded, autistic (James Dickey's novel, upon which Boorman's film closely relies, describes Lonnie as an albino boy with 'pink eyes like a white rabbit's' (Dickey 1970: 68) and a severe, sinister squint). In Barbara Creed's (1995) taxonomy of monster-bodies in horror films, he must sit alongside other 'infant body' monsters in movies like *It's Alive, Eraserhead* and *Basket Case*. Watched by his companions, Drew plays an amazing 'duelling banjos' routine with Lonnie; the tune serves as 'a leitmotif' for the four 'as their sense of security is eroded by both the current of the river and that of the narrative' (Ciment 1986: 126).

The men start their adventure, and their attention focuses on the river, on surviving the rigours of nature. They ride rapids, catch fish, make fires. Only Ed's failure to kill a deer mars their enjoyment of frontier-like existence. They are succeeding, even in this alien wilderness. But their enjoyment soon drains away, and the real horrors begin, when Ed and Bobby encounter two mountain men on the river bank (their vivid description by Dickey (1970: 115) matches well their on-screen menace). They try to pass pleasantries, but the conversation soon turns sour and the hillbillies rape Bobby at gunpoint, forcing him to squeal like a sow. As they turn on Ed, Lewis shoots one with his bow and arrow (note the weapons technology inversion: the yokels have guns, the city folk have knives and bows and arrows). The other hick disappears with his shotgun back into the woods. Slowly the four men try to make sense of what has happened in what Williamson (1995: 157) eloquently calls this 'rhododen-dron hell'; they first decide to head back, with the hillbilly's body, but they soon decide that any local law enforcers are sure to be against them (because everyone is related in such tight-knit communities), so they bury the corpse in a shallow grave, regain some of their composure and try to continue their quest.

The sequence following this only intensifies their nightmare. First Drew, for some reason not wearing his lifejacket, suddenly stops paddling, keels over in the canoe, and falls into the river. General panic ensues as his canoe (now containing only Ed) goes out of control, collides with Bobby and Lewis's, and all the men end up body-surfing down the rapids; Lewis breaks his leg in the process. Drew's body has vanished, and the others end up in a quiet pool downriver. Lewis says that Drew has been shot – by the toothless hillbilly from the rape scene. Ed decides he is still around, waiting to shoot the rest of them, unless he can kill him first. He climbs the steep river cliff face as night falls; he is clearly at the limit of his capabilities, and only just makes it. The next morning he wakes to see a bearded man with a shotgun in the distance. In a scene directly echoing the deer hunt, he raises his bow, and then – just as the man spots Ed and levels his shotgun at him – he shoots him dead with an arrow through the throat. Ed rushes to inspect the corpse and finds that it is not the toothless yokel from earlier. But did *this* man shoot Drew? Or has Ed killed an innocent man? Ed and Bobby weight and sink the corpse.

At this point, the film's horrors become interiorised, and for the rest of the time we are most aware of the horror in Ed's mind: the horror of having killed. Later, when they find Drew's broken body downstream, they struggle in vain to find definitive gunshot wounds (in Dickey's novel, Ed compares his skull wound with memories of news images of Kennedy's assassination); having completely lost themselves in their nightmare (Bobby says at this point, 'There's no end to it'), and without the injured Lewis's survivalist leadership, they weight and sink Drew's body too, decide on their story, and drift on downriver.

The last segment of the film finds the men at their destination, struggling to hide their 'crimes' while notifying the authorities of Drew's drowning. In realisation of their fears, it turns out that a local lawman's brother-in-law has gone missing on a lone hunting trip, and he is convinced that Lewis, Bobby and Ed know something; their alibi is weak and riddled with contradiction. The river is dragged, but no bodies are found; the river keeps its secret and the men are finally released. As they drive home, we see images symbolic of the impending end of the local landscape beneath the power company's floodtide: the local cemetery is being dug up, its coffins to be relocated, and the entire timber church is moved out on the back of a lorry. The men return home, and try to regain a grasp on 'normality'. In the final scene we are back at the river; a corpse rises to the surface – then we see Ed's startled, sweat-beaded face wake from this particular nightmare, see him in bed with his wife, see the look of horror and torment in his eyes.

Clearly, then, *Deliverance* works its source material in very different ways from *Texas Chainsaw Massacre*. But it still deals with similar rural-horror themes. As Michel Ciment (1986: 124) writes,

> the director turns a number of American myths on their head. The Garden of Eden is a poisonous jungle; and if Lewis is thinking of the euphoric adventurism of the frontiersmen... he is soon to be disabused: his wish to improve his 'image' among his social equals – an archetypal American trait – ends in total, devastating ignominy.

In the first section of James Dickey's novel – largely erased from Boorman's film version – Lewis talks about his survivalist philosophy: 'I think the machines are going to fail, the political systems are going to fail, and a few men are going to take to the hills and start over' (Dickey 1970: 51), building on this by talking about his bomb shelter and preparations for the aftermath of nuclear holocaust. (The same sense of apocalyptic fore-boding hangs over *Texas Chainsaw Massacre*, while in films like *The Hills Have Eyes*, radiation is explicitly named as the cause of rural degeneracy.) Ed, who narrates the novel, thinks about Lewis's comfortable middle-class suburban life. And, as the story of their lost weekend gets under way, we are made very aware of the limitations of Lewis's modern frontiermanship (in interview with Ciment, John Boorman says that Lewis breaking his leg reveals just how artificial and forced his 'relationship' with the wilderness is). Boorman sees the men's journey as a journey through American history:

> When they arrive at the village, they come into contact with people who live by the old frontier values, in an autonomous society in which they themselves build their houses, cultivate their land and defend themselves against outsiders, yet who are at the same time degenerates.
>
> (quoted in Ciment 1986: 129)

Ed's killing of the hunter, Ciment says, overturns the American myth of regeneration through violence; instead of giving strength, the act fractures the men's lives with the horror of the 'elemental impulses' of civilisation (Current 1986: 127). Here, in its reflections on violence, *Deliverance* echoes some of the themes of *Texas Chainsaw Massacre* (a pessimistic senseless-ness, for example), but also diverges from the sadistic brutalism of Hooper's film. In *Deliverance*, the 'innocent' are corrupted by their own violence in a way they are not in *Texas Chainsaw Massacre* (although in its sequel Stretch slaughters Leatherface and Choptop – but we never see her sub-sequent cold-sweating nightmares), and the horror is more internal, more human (its referents include, obviously, Vietnam). It is also reacting to and reworking an earlier dominant movie trope – the settler Western (with 'redskins' becoming rednecks) – but adding a tinge of conscience. As Clover (1992: 164) points out, 'the city approaches the country guilty', with the burden of economic exploitation, rural disenfranchisement and environmental recklessness also the cause of sweaty nightmares.

In the final section of this chapter, adapting a phrase from Creed's *Monstrous-Feminine* (1993), I will attempt to draw out some of the thematic issues from these films both as popular cultural products reflecting pre-vailing social fears, and as commentaries on the ambiguous place of the countryside in the popular imagination.

CONCLUSION: THE MONSTROUS-RURAL

Christopher Sharrett (1984: 255) states that the horror film 'has become heavily involved in asking questions about the fundamental validity of the American civilizing process'; he also notes the resonance between horror narratives and other critical assessments of American culture, especially those considering an inevitable, entropic implosion as the last self-immol-ating spasm of frontierist expansionism. The metaphorical time travel back through American history, which I would argue occurs as much in *Texas*

Chainsaw Massacre as it does in *Deliverance*, delivers its moral through relentless, nightmarish violence and horror; these are the hidden underside of America, and it is via media like the horror film that this gets worked through, not in the hope of redemption, but as a commentary on the 'fragile nature of all limits and all boundaries' (Creed 1995: 157) and even, as Wood (1984) suggests, a self-condemnation in which disintegration is ambivalently celebrated.

In an interesting survey of the iconography of small-town America in the movies, Emanuel Levy (1991) charts a move away from positive, romantic images of rural communities towards images of fragmentation and backwardness. In the 1970s, he notes, the film genre of horror began to replace the melodramas and comedies of previous decades. Indeed, from the 1950s onwards, small-town movies began to depict their settings as suffocating and repressive. As the huge upheavals of the 1960s were worked through in the cinematic medium, small towns and the countryside were increasingly portrayed as sites of contestation and decay, often embodied in those characters living there: 'The motifs of disenchantment and disintegration were expressed in individuals' loss of control. . . . Protagonists of small-town films have gradually lost their sense of identity and worth' (Levy 1991: 257).

Finally, the narrative of small-town movies, into which category films like *Deliverance* and *Texas Chainsaw Massacre* clearly fit (although Levy shies from citing the latter), collapses into 'nostalgia, paranoia and revenge' (Levy 1991: 257). In the liminal spaces of the Sawyers' Texas farm or a river in Georgia's backwoods, both characters and plots narrate these themes; there is a 'delicate balance between the bizarre and the ordinary' and always 'the ever-present threat of chaos' (ibid.: 260) in these portrayals of rural cultures. Here are societies which seem idyllic, but which are malignant, at a dead end, and viewed in the grip of their own death throes. That condition is the real horror of the rural.

ACKNOWLEDGEMENTS

Thanks to Tim Edensor and Ruth Holliday for chatting about the horrors of the rural. Special thanks to Nicholas Barker for loaning me a tape of *Deliverance*.

NOTES

1 I use the term 'monster' here as shorthand for the perpetrators of the horrors – who are usually portrayed as monstrous not just for their crimes, but in terms of their vile bodies and sick minds.

2 It should be noted that Christopher Sharrett (1984) draws a distinction in *Texas Chainsaw Massacre* between the figure of the senile drunk who offers warning comments on the horrors awaiting the film's victims – he represents the 'simple

rural folk' possessed of special insight and homespun wisdom – and the horrific amorality of the massacring cannibal Sawyer family featured in the film; likewise, in *Friday the 13th*, the warnings of a benevolent 'village idiot' go unheard by Annie, a soon-to-be victim (the 'idiot' is himself murdered in the film's first sequel, negating any salvation afforded those with special powers of insight; see Crane 1994).

3 This is one instance of camp comedy common in slashers; for example, the family in *Texas Chainsaw Massacre* is called the Sawyers (saw-yer). The theme of pulling commuters off the highway to meet their deaths in mad villages is also played out in the equally extreme 1974 Australian slasher *The Cars That Ate Paris*, which has a demolition derby climax, with the villagers destroying their village with their customised killer cars.

4 In a couple of the nine *Friday the 13th* films, 'Jasonism' becomes contagious, and any character in the film can become him and do his deeds.

REFERENCES

Bell, D. and Valentine, G. (1995) 'Queer country: rural lesbian and gay lives', *Journal of Rural Studies*, 11: 113–22.

Bunce, M. (1994) *The Countryside Ideal: Anglo-American Images of Landscape*, London: Routledge.

Ciment, M. (1986) *John Boorman*, London: Faber and Faber.

Cloke, P. and Milbourne, P. (1992) 'Deprivation and lifestyles in rural Wales – II. Rurality and the cultural dimension', *Journal of Rural Studies*, 8: 359–71.

Clover, C. J. (1989) 'Her body, himself: gender in the slasher film', in J. Donald (ed.) *Fantasy and the Cinema*, London: British Film Institute, 91–133.

—— (1992) *Men, Women, and Chainsaws: Gender in the Modern Horror Film* Princeton: Princeton University Press.

Crane, J. L. (1994) *Terror and Everyday Life: Singular Moments in the History of the Horror Film*, London: Sage.

Creed, B. (1993) *The Monstrous-Feminine*, New York: Routledge.

—— (1995) 'Horror and the carnivalesque: the body-monstrous', in L. Devereux and R. Hillman (eds) *Fields of Vision: Essays in Film Studies, Visual Anthropology, and Photography*, Berkeley: University of California Press, 127–59.

Dickey, J. (1970) *Deliverance*, Boston: Houghton Mifflin.

Finlay, M. (1995) 'The horrified position: an ethics grounded in the affective interest in the unitary body as psyche/soma', *Body and Society*, 1: 25–64.

Gillespie, A. K. (1984) 'A wilderness in the megalopolis: foodways in the Pine Barrens of New Jersey', in L. K. Brown and K. Mussell (eds) *Ethnic and Regional Foodways in the United States: The Performance of Group Identity*, Knoxville: University of Tennessee Press, 145–68.

Giroux, H. A. (1994) *Disturbing Pleasures: Learning Popular Culture*, New York: Routledge.

Glenn, J. (1995) 'Americana', *Observer* Preview section, 31 September: 46.

Grant, B. K (ed.) (1984) *Planks of Reason: Essays on the Horror Film*, Metuchen, NJ: Scarecrow Press.

Gutierrez, C. P. (1984) 'The social and symbolic uses of ethnic/regional foodways: Cajuns and crawfish in south Louisiana', in L. K. Brown and K. Mussell (eds) *Ethnic and Regional Foodways in the United States: The Performance of Group Identity*, Knoxville: University of Tennessee Press, 169–82.

Halfacree, K. (1996) 'Out of place in the country: travellers and the "rural idyll" ', *Antipode*, 28: 42–72.

Kinsey, A., Pomeroy, W. and Martin, C. (1948) *Sexual Behavior in the Human Male*, Philadelphia: W. B. Saunders.

Lesy, M. (1973) *Wisconsin Death Trip*, New York: Pantheon (reprinted 1991, New York: Anchor).

Levy, E. (1991) *Small-Town America in Film: The Decline and Fall of Community*, New York: Continuum.

Modleski, T. (1986) 'The terror of pleasure: the contemporary horror film and postmodern theory', in T. Modleski (ed.) *Studies in Entertainment: Critical Approaches to Mass Culture*, Bloomington: University of Indiana Press.

Philo, C. (1992) 'Neglected rural geographies: a review', *Journal of Rural Studies*, 8: 193–207.

Rubenstein W. D. (1981) 'New men of wealth and the purchase of land in nineteenth-century Britain', *Past and Present*, 92: 125–47.

Sconce, J. (1995) ' "Trashing" the academy: taste, excess, and the emerging politics of cinematic style', *Screen*, 36: 371–93.

Sharrett, C. (1984) 'The idea of the apocalypse in *The Texas Chainsaw Massacre*', in B. K. Grant (ed.) *Planks of Reason: Essays on the Horror Film*, Metuchen, NJ: Scarecrow Press, 255–76.

Urry, J. (1995) *Consuming Places*, London: Routledge.

Williamson, J. W. (1995) *Hillbillyland*, Chapter Hill, NC: University of North Carolina Press.

Wood, R. (1984) 'An introduction to the America horror film', in B. K. Grant (ed.) *Planks of Reason: Essays on the Horror Film*, Metuchen, NJ: Scarecrow Press, 164–200.

6

MAKING SPACE

Lesbian separatist communities in the United States

Gill Valentine

INTRODUCTION

The rural idyll is a concept which dominates both academic and popular understandings of the rural. Yet as Little and Austin argue,

> Despite its wide use in the past, the rural idyll as a concept, or set of concepts, has never been adequately unpacked. The term has been used to describe the positive images surrounding many aspects of the rural lifestyle, community and landscape, reinforcing at its simplest, healthy, peaceful secure and prosperous representations of rurality. Many writers have referred to qualities or attributes felt to be important to the rural idyll (Williams 1973; Short 1991; Laing 1992; Mingay 1989) but few have looked at any depth at how these may vary between groups and individuals.
>
> (Little and Austin 1996: 101)

Philo (1992) is one of several writers to promote a more postmodern approach to rural studies, urging rural geographers to be more sensitive to 'difference', specifically the experiences of marginalised 'others' such as women, children, the elderly, and lesbians and gay men within rural 'communities'. As part of this postmodern or 'cultural' turn in geography, a new body of work is beginning to emerge which is re-examining 'the rural' from different standpoints (Little 1996; Kramer 1995; Bell and Valentine 1995). This chapter focuses on the attempts of US lesbian feminists in the 1970s and early 1980s to produce their own very different sort of 'rural idyll' – non-heteropatriarchal space – through the spatial strategy of separatism. By constructing the rural as an escape from the 'man-made' city these women draw upon stereotypical representations of the rural as a healthy, simple, peaceful, safe place to live while also imagining their 'rural idyll' in a very different (and very politicised) way from traditional white middle-class understandings of rurality. Like traditional visions of rural 'community', many lesbian separatist imaginings of a common lifestyle also seem to have led to the marginalisation and exclusion of 'others'. The second half of the chapter therefore explores some of the tensions that arose in specific 'lesbian lands' as a result of attempts to create idyllic 'communities' by privileging the women's shared identities or sense of sameness as lesbians, over their differences.

Like much women's history, there is limited material recording lesbian separatist 'communities'. One, perhaps the only, attempt to document them is *Lesbian Land* by Joyce Cheney (1985). This is a collection of chapters each telling the story of a different womyn's project. Cheney began collecting material (letters, flyers and so on) on women's 'communities' in 1976 while living and working in one. In 1981, two years after her 'community' dissolved, she set out to travel around other 'lesbian lands' in order to record their experiences, and she organised two Lesbians on Land workshops. Cheney's (1985) research, carried out in the early 1980s, provides the main source of information for this chapter. The US focus of the chapter reflects both the richness of Cheney's material and also the absence of similar work from the UK. The issues raised are, however, relevant to the broader questions of sexuality and marginality in rural communities within which this chapter is framed. While the specific experiences of lesbian women reported here cannot simply be translated to the British context, they do serve to highlight the extent to which some sections of the gay and lesbian community feel excluded from mainstream rural lifestyles and culture. The particular responses of the US lesbians to this sense of marginality – the construction of alternative, separate communities – may not have been mirrored in Britain, yet they are worth considering here as providing an insight into the contestation of rurality, community and sexuality and illustrating the complex fractions that occur *within* identified groups.

LESBIAN LANDS: SPATIAL STRATEGIES OF SEPARATISM

In the 1970s some radical feminists began to identify heterosexuality as the root of all women's (lesbian, bisexual and heterosexual) oppression and to argue that separating from heteropatriarchal society was the only possible response. According to Faderman,

> Radical feminism propounded the behaviourist view of sexuality: as in a utopian socialist society where the individual could be conditioned to be non-violent, non-competitive, incorruptible, so too could women be conditioned to change their attitudes and desires. They could exit from the patriarchy through severing their relationships with men, which were seen as the cornerstone of the subordination of women, and they could learn not only how to make

a new society with women, but also how to respond sexually to
women.

(Faderman 1991: 207).

In other words, heterosexual desire, like the institution of heterosexuality
itself, was constructed and hence could be 'undone'. Feminism was the
theory, lesbianism the practice (Abbott and Love 1972; Brown 1976). In
this way, lesbian feminism tried both to fix lesbians as a stable minority
group and to liberate the lesbian in every woman by encouraging women
through consciousness-raising groups to choose not to be heterosexual.
'Feminism provided the ideological glue which wedded these two sometimes contradictory impulses' (Stein 1992: 38).

In order to avoid maintaining or perpetuating patriarchy in any way and
to enable women to construct an ideal new society beyond the influence of
men, some lesbian feminists adopted the spatial strategy of distancing
themselves from mainstream society by establishing separatist communities
that excluded all heterosexual and gay men. Although some women-only
communities were established in urban areas, for example in Toronto, the
aim of separatism was seen as best fulfilled in rural areas – because spatial
isolation meant that it was easier for women to be self-sufficient and purer
in their practices in the country than in the city, and because essentialist
notions about women's closeness to nature meant that the countryside was
identified as a female space. In contrast, the 'man-made' city was blamed
for draining women's energy. In this way women idealised the rural in a
political way – imagining it as simple, peaceful, safe space untainted by
patriarchy. This philosophy was reflected in women's fiction of this period
(Faderman 1991). For example, in *Demeter Flower* by Rochelle Singer
(1980), male civilisation is destroyed by nature, allowing women a new
beginning. And in Sally Gearhart's *The Wanderground* (1985), nature protects women, giving them the freedom to wander freely. In the 1970s,
therefore, 'many separatists established communal farms and became, as
one of their 1970s journals called them, country women' (Faderman 1991:
238). This was an important social movement at the time, and led to the
establishment of a whole circuit of communal farms or 'lesbian lands' in
the USA (Faderman 1991; Bell and Valentine 1995).

Separatists established land trusts to make land available to women for
ever. This control of space, they believed, was essential because it would
give women the freedom to articulate a lesbian feminist identity, to create
new ways of living and to work out new ways of relating to the environment, as these quotations (taken from Cheney 1985) from lesbian land
residents testify:

'It feels very real in the city, not having enough women's space to
go to. It's a constant struggle to have a space that we control.'
(Resident of A Woman's Place, in Cheney 1985: 23).

111

'We view our maintaining lesbian space and protecting these acres from the rape of man and his chemicals as a political act of active resistance. Struggling with each other to work through our patriarchal conditioning, and attempting to work and live together in harmony with each other and nature.'
(Resident of Wisconsin Womyn'’s Land Cooperative, in Cheney 1985: 132)

'Best of all, there is time and space for a total renewal of ourselves, our connections to each other and our earth.'
(Resident of Wisconsin Womyn's Land Cooperative, in Cheney 1985: 132)

These rural communities attempted to establish a lesbian feminist society in many different ways. Each looked to foster non-hierarchical ways of organising themselves and to be self-sufficient by developing their own economic institutions and skills, so that there was no need to go back into patriarchal society. Much energy was put into building new forms of dwelling and relearning old skills, such as fire-making, herbal medicine, and other survival skills. The communities also sought to articulate their identity by developing a women's culture (in terms of language, music, books and herstories). In this way they constructed very politicised visions of a 'rural idyll'.

Many of these lesbian lands also had a strong spiritual dimension. In keeping with their identification with nature, many of the communities celebrated the full moon, equinox, solstice and candlemas with ritual circles and used astrology and tarot readings (Warren 1980). As Faderman explains,

> Their idealised models were those ancient cultures, whether in myth or reality, in which women held secular power along with religious power. Lesbian-feminist spirituality was to resurrect the matriarchy, which would eliminate all of the destructive institutions of patriarchy – economic, political, sexual, educational, and return society to the maternal principle in which life is nurtured.
>
> (Faderman 1991: 227).

In particular, the emphasis was on goddess worship and on witchcraft, both traditions that were identified as women-centred and as symbolising resistance against misogyny and patriarchy (Ruether 1975; Ehrenreich and English 1973). In this way, 'Separate feminist spirituality is not only linked to a personal search for meaning and greater inwardness but is often closely connected with the acceptance of social responsibility and political activism' (King 1993: 116). This is reflected in these quotations from residents of lesbian lands (cited in Cheney 1985):

'We chant and sometimes share the names of our matrilineage. "I am Maryann, daughter of Mary, daughter of Mamie, daughter of MaryJane, daughter of Mary" '.

(Resident of A Woman's Place, in Cheney 1985: 21)

'We do have a basic ritual. We open the circle with a blessing and a purification, similar to Dianic. We begin with the oldest to the youngest. Lots of the time we use the salt water purification. Each one cleanses the other woman and takes away her negative energy.'

(Resident of the Pagoda, in Cheney 1985: 113)

'Each spring we were directed to do a medicine walk upon our boundaries. We walk 129 acres up and down the sides of the mountain.... This encircles us and protects ourselves, our animals, our community, our children from any interference.'

(Resident of Arco Idris, Cheney in 1985: 31)

Separatist feminism therefore 'became synonymous with the creation of a woman-identified Jerusalem' based on principles of 'sharing a rich inner life, bonding against male tyranny and giving and receiving practical and political support' (Ross 1990: 75). Faderman (1991) argues that these attempts to create new ideal ways of living represented a coherent philo-sophy that was not only challenging sexism and homophobia but also trying to create a unified alternative. She argues that it was intended that these

communities would eventually be built into a strong Lesbian Nation that would exist not necessarily as a geographical entity but as a state of mind and that might even be powerful enough, through its example, to divert the country and the world from their dangerous course.

(Faderman 1991: 217)

Despite their sharing of a vision of a 'rural idyll' out of the reach of patriarchy and promoting a shared lesbian feminist identity through their common goals, Cheney's (1985) research suggests that separatist communes of the late 1970s and early 1980s actually established very diverse ways of living. Each lesbian land was a very specific place. Attempts to organise a common lifestyle and to stake out a 'collective identity' by promoting a sense of sameness or unity around the identities lesbian and feminist often exposed differences between women which were negotiated and contested differently in different places. The second part of this chapter therefore explores how particular separatist groups attempted to manage their differences within their 'communities'.

LESBIAN LANDS AS SITES OF DIFFERENCE

Attempts to establish idyllic new ways of living required the women involved to define new ways of being and behaving. But this process of mutual identification automatically helped to generate homogeneity and to produce borders and exclusions (Young 1990). Perhaps the most significant debate faced by many separatists was how to define the sexuality of the rural space they were trying to create. It is generally assumed that separatism meant not only that lesbianism was 'destigmatised' among radical feminists but also that it was 'aristocraticised' (Faderman 1991). For some of the lesbian lands, this was certainly true. A desire for mutual identification or unity as lesbians led to the deliberate exclusion of heterosexual women and women with boy children because it was argued that these women were implicitly contributing to the maintenance of patriarchy. Other lesbian lands wanted to be open to all women, recognising that sexual identities are fluid and that many women go through stages of trying to redefine their sexuality, and therefore welcomed heterosexual or bisexual women as long as they did not bring men with them.

Boy children were a major stumbling-block for many of the lesbian lands. Ross (1990) documents how women with boy children were excluded from or made to feel unwelcome in the Canadian separatist community LOOT (Lesbian Organization of Toronto). Similarly, within US lesbian lands boy children were an important point of political cleavage. In one land, Kvindelandet, women with boys were turned away and the community even got rid of male foals and roosters in an attempt to create a 'pure' women-only space (Cheney 1985). But in other lands, the exclusion of women with sons was actively contested:

> 'Although some questioned boys' rights to be at OWL Farm, others of us defended them. It was open women's land, to exclude women because they had boy children would not have been "open". Many of us called ourselves separatists (others didn't) but our separatism never went so far as to exclude boy children.'
>
> (Resident of Nozoma Tribe, in Cheney 1985: 160)

Arco Iris allowed boy children and took a non-essentialist view of identity. As one woman explained,

> 'If a male child grows here and respects our ways and honours our ways and decides that he wants to stay here with us, at that time it would be taken before the womyn's counsel and decided upon. In my heart I feel that if that male child grew here and learned from us and our ways and decided he wanted to stay here, we would want him to stay here and be part of our community.'
>
> (Resident of Arco Idris, in Cheney 1985: 38)

In Maud's Land this was a fundamental debate through which the identity of the community was shaped:

'The male child issue has been the most painful one. We've all chosen to live without men, yet there are women now who are talking about having babies. This one has created not the most anger between us, but the most pain. We really have tried very hard not to go at each other. A lot of that anger or attack comes from fear; if she gets what she wants that means I may not be able to come here.'

(Resident of Maud's Land, in Cheney 1985: 88)

As this latter quotation implies, the collective identities of many of these separatist groups were not always stable but rather were fluid, as women contested and (re)negotiated their mutuality and consequently the boundaries of their 'communities'. In particular, many of the lesbian lands adopted changing positions towards monogamy and non-monogamy, experimenting with different ways of relating to each other. Redbird was one land documented by Cheney (1985) that (re)negotiated its identity around this issue, as this inhabitant explains:

'We reasoned that one falls in love because of a lot of conditioning (e.g. tall and slender), that everyone is loveable, and that if one focuses on the specialness of each person, one can still love anyone. So, we decided to choose lovers by drawing names out of a hat, and then go about loving that person, until, after several months, we'd re-draw and rearrange. I wouldn't recommend it. We tried. Oh we tried. Some combinations were just too hard, and we rearranged pairings. We were practising serial monogamy, one lover at a time. We thought we'd smash monogamy too by rotating through everyone in the collective until we had been with everyone and then having open sexual options within the collective.'

(Resident of Redbird, in Cheney 1985: 120)

Such attempts to determine new ways of living and a sense of sameness or togetherness often produced tensions between women living on the land who made different choices. In particular, celibate women often felt marginalised by women who were able to mobilise their lovers to support them in 'community' disputes. Others found it difficult to deal with relationships that had ended. Maximising a non-monogamous lesbian identity therefore often resulted not in unity, but in draining divisions and exclusions:

'Non-monogamy was politically correct. Monogamous couples felt an undercurrent of criticism of their relationships. Many women had several lovers on the land. It was difficult to get away from relationships one might not want to witness. Sometimes dealing with our feelings around our multiple relationships took so much energy that we had little left for anything else.'

(Resident of Nozoma tribe, in Cheney 1985: 152)

Class, a charged issue in the lesbian community, also took on larger dimensions in lesbian land environments. The land, while being a focus for unity, was also a source of division. Issues of ownership/co-ownership of sizeable investments such as the land itself, dwellings and 'improvements' (wells, roads, electricity, fences, outbuildings) often exposed differences between women. In many cases the land on which communities were established was acquired with inherited money or bought by a few women with well-paid jobs or savings who were able to put down cash or take out a mortgage. In most cases, the 'owners' would therefore have greater say in the direction of the community, with other women paying rent. The Arf community was one lesbian land organised in this way, as this woman explains:

> 'There are two collectives, the living collective and the legal collective. The legal collective is the women who are on the deed. They have complete say over everything that has to do with the deed and the taxes, paying the money, how the money gets spent, and who else gets put on the deed. The living collective has complete say over what happens living here and how we work that out. That is made up of whosoever is living here.'

(Resident of Arf, in Cheney 1985: 14)

Other lesbian lands handled the issues of class and ownership differently. As Faderman (1991: 237) argues, at this time, 'working class was seen as superior to the middle class, at least partly by virtue of its poverty, which attested to its moral innocence in a corrupt society'. Middle-class women bore a heavy burden of guilt. As a resident of Nozama Tribe recalls, 'we played "more oppressed than you" '. She explains how women manipulated each other to get what they wanted: 'whoever hurt us was judged as "middle class" and condemned' (Cheney 1985: 162). Another recalls some of the pressures this led to:

> 'One woman among us had a large inheritance that she was trying to deal with in a class-conscious way.... Eventually we decided to split the money 3 ways – $50,000 each to city women, country women and women of colour ... we gave a small amount of money to the Oregon's Women's Land Trust. We could have completely paid off OWL farm with some of the money but didn't because of our white middle-class guilt and pressure from other women that we should not use the money for "ourselves".'

(Resident of Nozama Tribe, in Cheney 1985: 160)

116

While class was negotiated and contested within communities, lesbian lands also often came under pressure from outsiders who did not respect the boundaries of the communities and accused the residents of class privilege. One woman describes a bitter dispute her community had with a group of travelling women who tried to move in and appropriate the commune, claiming that 'all property was theft':

'There were some women who felt that we were elitist, capitalist, racist, classist snobs, taking our privilege with private land; they didn't respect private land.'

(Resident of Arf, in Cheney 1985: 15)

Claims of racism and a lack of tolerance of disabled women were other fissures of difference that split the fragile unity of lesbian land communities. Stein (1992: 36) argues that 'Lesbian feminism and the women's liberation movement in general drew heavily upon the images and symbols of Black Power and shared its commitment to authenticity, redefining and affirming the self, and achieving individual recognition via group identification.' But despite this, few black women were involved with white women in establishing lesbian lands.

Faderman argues that one explanation for this was that black women 'felt greater solidarity with "progressive" minority men than with white lesbian-feminists who, it seemed to them, were denying that race could be as much a source of women's oppression as sex' (1991: 241). The evidence of Cheney's research (1985) is that many black and Jewish lesbian feminists felt marginalised from communes because of the communities' inherent whiteness and their lack of consciousness about the specificity of oppression – a criticism shared by disabled women, who often felt that lesbian lands were not set up for or would not respond to their needs. In particular, many of the communities emphasised a physical commitment to the land through shared physical labour. This emphasis on the 'body' meant that many disabled woman could not participate. Thus by attempting to mobilise a lesbian feminist identity, the women involved in establishing lesbian lands often privileged their sameness over their differences, as these quotations (cited in Cheney 1985) demonstrate:

'I do not want to define the Jewish struggle and I do not want to divide the womyn's movement into ethnic camps but I simply cannot sit by and let a group of white people reclaim my land. I also cannot sit silently by while white womyn compete with us for the little bit of validation we have received from the womyn's movement for being womyn of color. The Third World consciousness in the womyn's movement is relatively new, and the movement was dominated by white feminists for longer than we have enjoyed any validation for our struggle.'

(Resident of La Luz de la Lucha, in Cheney 1985: 69–71)

'One woman couldn't deal with seeing a disabled woman, because she had been disabled and didn't want that in her life unless I could somehow be happy. And I've not been at a place to do that.... There were those who would have preferred for me to take drugs in order for me to be more functional. I feel they had no caring for what happened to me tomorrow but only whether I was able to take care of myself and do physical work today. It's clearer to me now that I don't want to live with wimmin who have that kind of expectation of me, because it isn't what I choose to do with my life.... My rights as a disabled person to live on the land were not recognised... I wanted the country to be a healing place for me. The city had definitely been a place that made me more disabled. Golden was also a place that made me more disabled. The stress level and the constant pressure to push myself to the limits of my physical capacity have made my condition worse.'

(Resident of Golden, in Cheney 1985: 52)

While the emphasis on escaping patriarchy meant that many lesbian lands emphasised the residents' shared identities as women over their differences, not all separatist communities had the dominant aim of creating woman-centred space. Rather, some lesbian lands maximised other identities, with the aim of escaping disability oppression by making independent space for disabled lesbians, for example; others attempted to create non-racist environments. As one of the founders of Beechtree commune explains:

'I always had a dream of being in the country; I wanted to develop space for disabled people to live autonomously. There are a lot of care facilities of different levels, some apartments, but they all feel restrictive.... I began networking with disabled lesbians together, in an ableist-free environment. Maybe we wouldn't stay together forever, but we deserved to live validated lives. There's no place I can feel totally validated unless I'm with a lot of other disabled lesbians... I have a disability; my culture is different, my herstory is different and my reality is different.'

(Resident of Beechtree, in Cheney 1985: 42)

Other communes were established by native American women and by women of colour. Arco Iris was a community specifically for womyn of colour. White womyn were allowed to live there only if they came with a womyn of color or had bi-racial children, – or by special invitation of womyn in the community. Another example described in *Lesbian Land* is La Luz de la Lucha (Cheney 1985).

Thus, lesbian separatist attempts to establish 'idyllic' ways of living in the countryside appear to have unravelled because, in common with traditional white middle-class visions of 'rural community', attempts to create unity and common ways of living also produced boundaries and exclusions.

CONCLUSION

The focus within rural geography on white middle-class visions of a 'rural idyll' obscures the fact that 'other' groups have also idealised 'the rural' as a peaceful, safe place and sought to establish their own versions of 'community' life away from the city. Lesbian separatists are one example of a marginalised group who have idealised the rural and attempted to live out very politicised visions of a rural lifestyle, by emphasising their shared identities as women through the spatial strategy of separating from hetero-patriarchal society.

There does not, however, seem to have been one common vision of how to create lesbian feminist ways of living and lesbian feminist space. Each of the lesbian lands described by Cheney (1985) appears to have been a specific and unique place that defined its own vision of how women should live together and established its own collective identity. Some communities emphasised, for example, the creation of non-racist environments or space for the disabled to live independently; while others focused on issues of monogamy or self-sufficiency. As Stein (1992: 37) has argued, 'the lesbian-feminist movement consisted of hundreds of semi-autonomous, small-scale groups that were never centred'.

Each lesbian land, by defining its own common ways of living and appropriate ways of behaving, constructed its own shared identity or groupness. These desires for mutual identification or homogeneity simultaneously appear to have generated boundaries and exclusions. As Iris Marion Young (1990: 301) argues,

A woman in a feminist group that seeks to affirm her mutual identification will feel and be doubly excluded if by virtue of her being different in race, class, culture or sexuality she does not identify with the others nor they with her.

As Cheney's (1985) research shows, women felt excluded within lesbian lands for many reasons. Some refused to be controlled by the dominant identities of the communities they were living in and pushed against their boundaries by mobilising the performance of 'other' identities. Others chose to leave one community and join or set up another. In this way, the lesbian lands were not stable communities but were fluid, with new women coming and going as different identities were maximised and minimised.

As a resident of Golden (Cheney 1985) explained, this was often a negative process:

'It made me feel guilty all the time to gang up on wimin. Yet in both

119

cases, the wimin who were asked to leave were not fitting in with the existing group. This is a little bit of a different situation than I was in, where the wimin at the land basically split into two factions and one faction left. In the other cases it was one woman vs. the rest of the group. I don't know if that makes it any more justifiable. . . . I really encouraged a diversity, and thought diversity was something that would help us grow. Now I feel that, in a lot of ways, the diversity made it really hard. We were coming from so many different places, and had so many different goals, and were all so scarred as lesbians living in the patriarchy that we were bound to take some of that out on each other. Striving for good communication is about my first priority now.'

(Resident of Golden, in Cheney 1985: 54)

As this woman clearly articulates, while some lesbian lands wanted to recognise and value diversity among the inhabitants of their communities, the reality was that in many lands, identities were not equally valued, rather some were privileged over others. The tensions within these lands developed as a product of the inequalities of power and hierarchies that were constructed between women. In this way the space of separatism clearly exposes feminism's 'complex and divided self' (Rose 1993: 153). These processes of power undoubtedly also operate in traditional rural 'communities'. But whereas the politicised nature of lesbian separatist communities meant that women were aware of and reflected upon their 'divided selves', it is all too easy for conventional rural dwellers and indeed academics to ignore the processes of power which also operate in traditional imaginings of the rural idyll.

ACKNOWLEDGMENTS

Part of this chapter was originally published in James, J. P., Nast, H. and Roberts, S. (eds) *Thresholds in Feminist Geography*. It is reproduced in this volume by kind permission of Kouman & Littlefield. I wish to thank Sue Roberts, Heidi Nast and John Paul Jones for organising the excellent conference New Horizons in Feminist Geography (and the accompanying workshop sessions) at the University of Kentucky, Lexington, USA, where this chapter was originally presented. I am grateful to all those in my workshop sessions, especially the facilitator Audrey Kobayashi, for their insightful comments which helped me to develop this chapter. I also wish to thank David Bell for finding Cheney's book *Lesbian Land* and for his continued academic and personal support. Finally, I am grateful to Jo Little for her editorial guidance.

NOTE

1 Many lesbian separatist groups attempted to reclaim or redefine language as part of their attempt to create a women-centred culture, for example using words such as womyn, wimmin, herstory. In the quotations cited in this paper I have used the spellings adopted by the original author and consequently there is no consistent use of terms within the paper.

REFERENCES

Abbott, S. and Love, B. (1972) *Sappho Is a Right-On Woman: A Liberated View of Lesbianism* (Stein and Day, New York).

Bell, D. and Valentine, G. (1995) 'Queer country: rural lesbian and gay lives', *Journal of Rural Studies* 11, 2: 113–22.

Brown, R. M. (1976) *A Plain Brown Wrapper* (Diana Press, Baltimore).

Cheney, J. (1985) *Lesbian Land* (Word Weavers, Minneapolis).

Ehrenreich, B. and English, D. (1973) *Witches, Midwives and Nurses: A History of Women Healers* (Writers and Readers Publishing Collective, London).

Faderman, L. (1991) *Odd Girls and Twilight Lovers: A History of Lesbian Life in Twentieth-Century America* (Penguin, Harmondsworth).

Gearhart, S. (1985) *The Wanderground: Stories of Hill Women* (Women's Press, London).

Harvey, D. (1993) 'Class relations, social justice and the politics of difference' in Keith, M. and Pile, S. (eds) *Place and the Politics of Identity* (Routledge, London) pp. 41–66.

Jackson, S. (1992) 'The amazing deconstructing woman', *Trouble and Strife* 25: 25–35.

King, U. (1993) *Women and Spirituality: Voices of Protest and promise* (Macmillan, Basingstoke).

Kramer J. L. (1995) 'Bachelor farmers and spinsters: lesbian and gay identity and community in rural North Dakota'. In Bell, D. and Valentine G. (eds) *Mapping Desire: Geographics of Sexualities* (Routledge, London).

Laing, S. (1992) 'Images of the rural in popular culture 1750-1990'. In Short, B. (ed.) *The English Rural Community: Image and Analysis* (Cambridge University Press, Cambridge).

Little, J. and Austin P. (1996) 'Women and the rural idyll', *Journal of Rural Studies*, 12, 2: 101-11.

Mingay, G. (ed.) (1989) *The Rural Idyll* (Routledge, London).

Rose, G. (1993) *Feminism and Geography: The Limits of Geographical Knowledge* (Polity Press, Oxford).

Ross, B. (1990) 'The house that Jill built: lesbian feminist organising in Toronto, 1976–1980' *Feminist Review* 35: 75–91.

Ruether, R. (1975) *New Woman, New Earth: Sexist Ideologies and Human Liberation* (Seabury Press, New York).

Short, J. (1991) *Imagined Country: Society, Culture and Environment* (Routledge, London).

Singer, R. (1980) *The Demeter Flower* (St Martin's Press, New York).

Snitow, A. (1990) 'A gender diary', in Hirsch, M. and Fox Keller, E. (eds) *Conflicts in Feminism* (Routledge, New York).

Stein, A. (1992) 'Sisters and queers: the decentering of lesbian feminism', *Socialist Review* 22, 1: 33–56.

Warren, M. A. (1980) *The Nature of Woman: An Encyclopedia and Guide to Literature* (Edgepress, Inverness, California).

Whisman, V. (1993) 'Identity crises: who is a lesbian anyway?' in Stein, A. (ed.) *Sisters, Sexperts, Queers* (Plume, London) pp. 47–60.

Williams, P. (1973) *The Country and the City* (Chatto and Windus, London).

Young, I. M. (1990) 'The ideal of community and the politics of difference', in Nicholson, L. (ed.) *Feminism/Postmodernism* (Routledge, London) pp. 300–23.

7

RURALITY AND 'CULTURES OF WOMANHOOD'

Domestic identities and moral order in village life

Annie Hughes

INTRODUCTION

It is only in the past five years that rural geographers have begun to recognise their neglect of multiple forms of 'otherness' in the contemporary countryside. Influenced by the pervasiveness of both poststructuralist and feminist critiques in social sciences, rural researchers are beginning to address issues of power and marginalisation in their writings, critiquing their traditional paradigms, which have turned a blind eye to the possibility of other human groupings in rural society (Philo 1992).

Feminist rural researchers have gone some way to recovering the experiences of rural women whose lives have been marginalised by mainstream analytical and empirical analyses. Feminist research has highlighted how inadequate conceptualisation of terms such as 'work' and 'the family' in conventional rural research has led to the neglect of rural women's voices in academic research. By far the largest body of work in feminist rural research has focused on women in farming, highlighting the importance of gender or, more particularly, patriarchal relations in structuring the farm labour process (see Whatmore 1990, for example). This research has highlighted the analytical constraints of traditional studies of the rural labour process which neglected the importance of what could be termed 'women's work' in the informal and reproductive economies. Importantly, this body of literature has placed reproductive spaces back on the rural research map and has highlighted the need for a reconceptualisation of the term 'work' to include the dynamics of the domestic and informal economies.

However, although there has been a heightened awareness of gender divisions in farm households and a recognition of the importance of reproductive labour in sustaining and restructuring the farm economy, far less attention has been paid to the broader experiences of *gender identities* within the rural context (Redclift and Whatmore 1990). It could be argued that the lived experiences of non-farming women have remained largely untheorised, marginalised not only from conventional rural research but also from feminist investigations into rural life and lifestyle. The need for broader research which focuses on the experiences of (different) women in the contemporary countryside has become salient as the rural social structure is transformed by whole sets of incoming groups not directly linked to the farming economy.

Drawing on work in rural studies concerning the social construction of rurality and recent feminist theories concerned with the fluidity and fragmented nature of gender identities, this chapter focuses on rural women's domestic experiences. In particular, it focuses on women's domestic identities in the context of changing rural spaces and assesses how these identities are informed by dominant constructions of rural femininity and the resulting moral orders of village life. Gender and rurality are, therefore, treated not as fixed, unchanging categories but as 'unstable and interactive reference points' (Whatmore *et al.* 1994: 4) constructed through social and cultural practices which have given them meaning in everyday life.

The aims of this chapter are threefold. First and foremost, it (re)values the domestic sphere as a legitimate realm of rural research, a realm all too often neglected and marginalised in mainstream analysis. In so doing, it recovers the lived experiences of women within this sphere and their particular associations with rural places. Secondly, it assesses the importance of 'the domestic' to the gender identities of rural women within the contexts of the socially constructed, and contested, categories of rurality and 'womanhood'. Finally, it investigates the ways in which rural women reinforce and contest dominant constructions of rural femininity as they negotiate their identity through the economic, social and moral orders of village life. By acknowledging the diversity of female domestic experiences, the chapter highlights the fragmented nature of rural femininity as it is cross-cut by other social identities such as class and age.

Recent qualitative research in Wales is drawn upon to investigate this threefold aim. By focusing on the domestic lives of women living in the contemporary countryside, this chapter centralises women's rural experiences and 'the domestic sphere' more generally in rural research.

FEMININITY, DOMESTICITY AND RURALITY: UNRAVELLING THE LINKS

A substantial body of literature in rural studies has begun to rethink and redefine rurality as dynamic and unstable social constructions rather than as fixed geographical entities (see, for example, Bunce 1994; Cloke and Milbourne 1992; Crouch 1992; Halfacree 1993; Mormont 1990). This work has begun to recognise the significance of the category 'rural' in everyday life as it is constructed through, and embedded in, historically and spatially specific social and cultural practices which have invested it with particular significance (Whatmore *et al.* 1994). Moreover, it is argued that the multiple meanings attaching to rural areas must be uncovered if contemporary rural experiences are to be fully understood and explained. Rurality is culturally defined and, as a result, the social, economic and cultural meanings inferred in relation to, and embedded in, rural places need to be addressed if we are to understand how these discourses inform contemporary experience.

Authors concerned with the meanings of rurality have highlighted the centrality of home and community to dominant constructions of the rural. As Davidoff *et al.* (1976) argue, 'The very core of the [rural] ideal was home in a rural village community.' Feminist researchers have become interested in the place that women have been assigned in dominant constructions of rurality, arguing that these constructions incorporate specific ideas about the proper roles for men and women (Little 1987). It is argued that women's 'natural' role was placed centrally within the home and the community, nurturing and caring for her family. A particular construction of rural femininity has developed, linking womanhood and domesticity with notions of the organic community, a construction that has been reproduced within a range of contemporary writings about the rural. Indeed, Nead (1988) argues that 'natural' gender differences were central to representations of rural idyll 'with women naturally natural mothers' (Rose 1993: 95). It has been suggested that these 'natural' gender differences have become formalised within (dominant) rural discourses as the 'domestic idyll' has retained its importance in contemporary rural ideology (Stebbing 1984; Little 1986, 1987). The prevalence of constructions of the 'domestic rural woman' is highlighted by Braithwaite (1994: 12), who argues that '[T]he stereotype of a rural woman is that of a family woman, traditional and conservative, absorbed in the care of the home.'

The need for an examination of the historically specific construction of gender identities in the rural context must be seen in the light of more general debates concerning the social construction of femininity (and masculinity) and the ways in which gendered identities are constituted in place. This work reflects wider concerns in feminist writings concerning differences among and between women, debates which disrupt the possibility of common gender identities (Nicholson 1990).[1] Gender identities and ideas of gender difference are not fixed, but rather are constantly changed and reworked. Femininities (and masculinities) are historically specific (Hall 1992) and embedded in particular spatial contexts (Moore 1988). The 'domestic woman' is not universal, and certainly not the 'natural' identity of women ('natural' meaning 'as decreed by nature'). The sexual division of labour is not rigid, not given in nature, but constructed in history and, as such, socially defined (Hall 1992). The meanings of manhood and womanhood are not inherent qualities linked to biological sex but identities continually being constituted, reproduced and contested at particular times in particular places (Glenn 1994). Moreover, the category 'womanhood' is at any one time, in any one place, cross-cut by myriad other social divisions such as class, age and race which intertwine to inform gendered identities.

If we are to accept that female (and male) experience does not conform to a uniform and stable concept of femininity (and masculinity), the importance of understanding the historically specific nature of gender identities and their spatial embeddedness seems paramount. The construction of the 'domestic woman' embedded in rural discourses is not woman's natural state, but constructed, and reinforced, through patriarchal relations. If we are to understand and explain the domestic experiences of rural women we need to take account of the ways women adhere to and contest dominant constructions of rural womanhood.

Feminist rural researchers are now beginning to embrace these debates to examine women's gendered experiences in the context of changing 'rural' spaces. However, this work, again, has focused on women in farming, leaving the experiences of women in the broader rural community untheorised. Brandth (1994), for example, examines the ways in which farm women construct their femininity in the context of changing agricultural practices. In particular, she investigates the meaning of femininity to women who routinely carry out tasks traditionally performed by men. Although she argues that femininity is being reconstructed to include these new tasks, she stresses the importance of domesticity to the feminine identity of her respondents. Femininity is, therefore, not cast adrift from the domestic qualities of the housewife. This chapter broadens the narrow focus on farm women to investigate how rurality and domesticity interlink to inform the gender identities of rural women more broadly.

In the remainder of this chapter the debates concerning the historically and spatially specific nature of femininity are drawn upon to investigate the gender identities of contemporary rural women. Recent research is employed to assess the complex relationships between femininity, domesticity and rurality through a discussion of the domestic identities of rural women. The importance of domesticity to women's sense of worth is examined in the context of the place-specific construction of the 'domestic rural woman'. The economic, social and moral orders of village life are investigated to show how dominant domestic identities are not only sustained, but also challenged and contested. This is achieved with reference to the interweaving social divisions of class, age and background as they intersect with the physical features of rural areas and the dominant constructions of femininity to influence women's domestic identities. This approach allows an understanding of how (different) women relate to, and contest, (dominant) constructions of womanhood in rural discourses and how women who do not 'fit' these constructions are marginalised in rural society.

126

OTHER RURALS – ADDING THE GENDER DIMENSION INTO RURAL RESEARCH

The findings drawn upon in this chapter are based on research carried out in 1994 in two villages, Ditton and Llangeley,[2] situated in mid-Wales close to the English border. This research adopted a feminist epistemological stance which highlights the subjective nature of knowledge,[3] to argue for the legitimation of women's subjective experiences as a valid way of knowing and understanding the rural. Not only did it aim to add women into rural geography but also it provided the theoretical basis to critique dominant agendas in rural geography which have tended to focus on non-gendered experiences, and academic discourses, of rurality. A qualitative method of enquiry was adopted as providing the best tools with which to investigate women's subjective experiences of living in the countryside.[4] Sixteen in-depth interviews were carried out in Ditton and thirteen in Llangeley as well as three group discussions. Both villages were located in a parish of only 534 people. The surrounding area, although still dominated by farming, has experienced in-migration during the 1980s and early 1990s, by predominantly (but not exclusively) of retired English couples.

This chapter draws on the information from only fifteen of the twenty-nine women interviewed, largely because of space constraints. However, other women interviewed shared the views expressed in this chapter. Although it is impossible to provide full biographical details of all the women, it is important to put their lives in some sort of context. Further biographical details can be found in the text. Phyllis and Catherine were both retired farmers' wives who had lived in the area all their lives, bringing up their now grown-up families on their farms. Rebecca, Anne and Louise represent the younger generation of farmers' wives, all of them having been brought up in the area. Emma and Rachel were younger non-farming women, both of whom had lived in the villages all their lives. Elaine, Pauline, Rose, Mary, Jenny and Lizzy were non-local women of varying ages who had moved into the villages later in their lives. All the women interviewed were married, most with children, reflecting the predominance of traditional family structures in the villages.

TIMES PAST (?): RURAL WOMEN AND TRADITIONAL DOMESTIC IDENTITIES

Traditionally in Ditton and Llangeley a woman's place was in the home. This was particularly the case in farming families, but not exclusively so. Phyllis, a retired farmer's wife and mother of six children who had been brought up in Ditton, recounted her experience:

'When I was brought up the woman was always working at home. There was no doubt about it. I can't remember any woman from a farm, or any other woman for that matter, who had a job. They were always at home.'

Phyllis argued that a rural woman participating in paid employment was 'unthinkable', 'a real oddity' and very 'strange'. This attitude was also reflected by Catherine, another retired farmer's wife, in her sixties, living in Ditton. She stated, 'When I was growing up we were always taught to cook and get meals on time because the men were the breadwinners and they needed a good dinner every day. That was the way it was.' Neither Catherine nor Phyllis had ever had a paid job, both accepting their place in the home and on the farm. Similar views to those expressed by Phyllis and Catherine were held by other farmers' wives living in Llangeley and Ditton. Rebecca, a farmer's wife in her fifties, commented:

> 'When I was younger all the farmers' wives around me were at home; they cleaned and that is all they did. They may go to market one day a week ... otherwise they were at home all the week doing the cleaning and the cooking.'

By all accounts it seemed that farmers' wives, and women living in the villages more generally, adopted a very traditional domestic role, staying at home and looking after their homes and families.[5]

The necessity of women's domestic role was viewed, by the women, in terms of the materiality of rural living. For example, Ditton did not receive mains electricity until 1952 and the outlying farms received it later still. By virtue of the fact that domestic chores were carried out by hand, housework was a full-time job. Rebecca pointed this out during her interview:

> 'When my kids were small ... there was no water, and there was no electric. With four kids you were all day carrying water, washing by hand, of course, and all the rest. So, therefore, your days were full of just working in the home. Now you've got washing machines, microwaves, Agas and everything is done so much quicker, isn't it.'

In reiterating this point Phyllis highlighted her average week as a young mother: 'We were always at home. We just didn't have time [to go out to work]. Monday was washing, Tuesday was churning and Wednesday was baking bread and cakes. I remember washing took hours. Everything took so long.'

Additionally, and in comparison with the situation today, women had fewer opportunities to drive, which made working outside the home more problematic. However, Anne, a farmer's wife in her fifties, suggested that the perceived duty of the rural woman was to stay at home: 'It was taken for granted that she [got] the meals.'

Paid employment undertaken by village women tended to be informal and part-time, and an extension of their domestic duties. It was fitted around their childcare commitments and domestic responsibilities, and tended to be near their homes, and certainly in the village. In arguing these points Rebecca highlighted her own situation: 'I was at home when I had my kids although I used to go to work at the pub in Llangeley.... I used to clean in the morning... but then I could take whoever was the baby with me.' Women who did partake in paid employment accepted their primary role as wives and mothers, and as such did not upset the moral order in the village, which placed women centrally within the home and the community. Although rural women were expected to stay at home it was the more affluent women who actually did so, while those women who were less well off worked informally around the village. This highlights interesting links between class, domesticity, womanhood and moral order. Women went out to work if it was financially necessary as long as they did not challenge the prime domestic, and therefore proper, identity of women.

Women's identity was centred around domesticity and the home. Indeed, proficiency in domestic tasks was still seen as important to the older women's sense of self-worth. In her interview Catherine took great delight in explaining how to make home-made pies, arguing, 'they're beautiful and... I tell you now after [tasting] those, you will never buy a pork pie again'. Phyllis also highlighted the importance of domestic tasks to her sense of worth, arguing that she found housework extremely fulfilling. She continued,

'I was brought up to make things... when you had to get down on your knees and scrub and polish the floor, it was jolly hard work... but you got a sense of real satisfaction.... You could have a sit down afterwards and feel proud.'

She argued that if she could do a bit of knitting or sewing she felt her day was complete.

REDEFINING FEMININITY: DOMESTICITY, RURALITY AND EMPLOYMENT

The majority of working-age women living in Ditton and Llangeley now, however, do undertake some kind of paid employment. Louise, a mother of two in her early forties, who had never worked, argued, 'I am in the minority now.' Elaine, a mother who had moved into Ditton fifteen years ago, agreed, claiming that rural women's lifestyles have changed, 'and they do not bake religiously every day'. She went on to comment, 'We are being thrown into the twentieth century.' However, as Louise pointed out, this had only occurred in the past ten years. By and large, the increase in the number of women going out to work was attributed to financial reasons although the availability of modern appliances in the home and the increasing numbers of women who had access to a car were also mentioned.

On the surface, it would appear that women were challenging the traditional order in rural society and rejecting their domestic identity. However, it became clear that the majority of women worked part-time and that their work tended to be fitted around their domestic responsibilities.[6] Childcare provision in the area was poor, as it is in many rural areas (Stone 1990). There were no registered child minders or playgroups in either Ditton or Llangeley where women could leave their children, although many women relied on informal childcare. Both Rebecca and Catherine looked after their grandchildren for short periods throughout the week, and Emma, a young mother brought up in the area, relied on her own mother if she required a baby-sitter. Louise commented, 'the majority of women in this area don't go out to work until their children are at school and then they find jobs that fit in with school hours.' In fact, Louise could not identify any mother who left her children with a formal child minder or at a crèche.

The general attitude prevailing was that it was a woman's duty to stay at home with her children, particularly when the offspring were younger. This was felt by both younger and older women, by incomers and locals, by the well off and less well off. When asked if she thought women should stay at home, Louise said, 'Yes, until they have finished at primary school.' (She qualified her answer by stating that, if the family were having financial problems and the woman had to go to work, then the situation would be different.) Indeed, Phyllis stated this attitude categorically:

'I think a woman's place was traditionally in the home and I am old-fashioned probably but I still think so... that is what has gone wrong with society. Women are going out to work and they are not doing their job at home looking after the children and the home.'

Incomers as well as the local women held this opinion. Jane, a young mother of three who had moved into the village fourteen years previously as a newly-wed, argued, 'I never wanted to go back to work when the children were at home.... I didn't have to go out to work so I thought it was important to be at home and I still think that.'

There are certainly more physical barriers for women wanting to go back to work in rural areas, compared with their urban counterparts. The lack of adequate public transport and childcare facilities – indeed, the lack of suitable employment – all intertwine to reduce women's employment opportunities. However, it was not only physical barriers that prevented women from taking on paid employment. Emma commented, 'I think... living in this area has restricted me in my job and in my outlook especially. You become cautious of pushing yourself.' Emma felt inhibited because it was not deemed acceptable for a young mother to obtain a full-time career. This attitude, of course, is influenced by more than the physical and social constraints of rural areas. Family expectations, personal motivation and confidence all enter the equation. However, what is interesting is that Emma felt that living in a rural area had narrowed her opportunities, not only as a result of the lack of employment opportunities, but also because of the fact that a woman's role was perceived to be centrally placed within the home. Indeed, the younger women said that they felt frowned upon by older women because they were 'out the house so much'.

Although it is acceptable for women to go out to work part-time, they are still expected to take on the primary care of their children and, indeed, carry out the majority of domestic tasks. Women who undertook paid employment without apparent financial necessity were deemed to be 'just gallivanting', and in small villages such as Ditton and Llangeley their actions were noted. Elaine felt that there was 'more pressure to conform to what [rural] society expects of you'. She went on to suggest that if women, particularly those who had moved into the area, did not try to fit in with the village, they would not be accepted. Any woman perceived as rejecting the traditional construction of femininity is more readily identifiable in small rural communities:

> 'You have to work hard to prove yourself to feel you belong. There is a lot of pressure, whereas in the city . . . you don't have to make such an effort. You have to conform to their expectations or they will gossip . . . about you.'

The social constraints on women are much more apparent in rural areas and if, as Elaine stated, 'you stray from the fold . . . you are noticed and judged'. Ditton and Llangeley are very small villages and, as Rachel pointed out, 'you generally know how people are fixed'. Although women were beginning to challenge the traditional life of 'the rural woman' by going out to work and gaining interests outside the home their domestic role has remained relatively unchanged. Most women suggested that their partners did relatively little to help around the house. Clearly, rural areas are being influenced by modern life and rural women are beginning to make more choices. However, the research shows clearly that rural women's choices are constrained not only by the physical features of the area but also through the moral and social orders in village life which place women within the home. These constraints are sustained through other women's attitudes to 'working women' and, perhaps more importantly (and subtly), through the social norms and expected behaviours in rural life.

THE CONTINUING IMPORTANCE OF DOMESTICITY TO RURAL FEMININITY

It became clear, as the research progressed, that the possession of domestic abilities was part and parcel of being deemed 'a rural woman'. Indeed, domesticity was held in very high regard in the village. Rose, an incomer of only a matter of months, marvelled at women she termed 'country ladies':

> 'Some of these country ladies, I just do not know how they do it. . . . One particular lady, she cooks, and if there is something going out in the village she will pull her weight and she will bake her few cakes and she will, you know, always do her bit to help. She is a wonderful cook, a real countrywoman.'

Mary, a woman who had moved into the area two years prior to our conversation, argued:

> 'Even though most of them [rural women] work they are very talented. . . . They are marvellous cooks and I have learned a lot from them, it amazes me. . . . I don't know too many that don't work even if it is only the odd thing . . . but they bake a lot and they knit and they do tapestry . . . they are very talented.'

Note the comment, 'Even though [they] work, they are very talented.' She is suggesting that although rural women go out to work their true talents lie in the home. She continued, 'I don't think they [rural women] have changed as much as in the town. Take Catherine . . . I can't see her changing. She is a very solid good woman . . . she is always baking.' Far from being seen as menial, domesticity was praised both by local women and those who had moved into the village.

It could be argued, however, that there is another, more subtle, layer to Mary's discussion of 'rural women'. Interestingly, Mary does not include herself as 'one of them' even though she had lived in a rural area for extended periods of her life. When asked about the lives of rural women, she always referred to 'them' rather than 'us'. Elaine maintained a similar attitude, stating, 'There still are rural women. . . . The older generation . . . some of them still live very quiet lives at home, the old countrywomen.' Although Jenny argued that images of rural women baking their own bread and churning their butter went out with the Ark, she suggested that real countrywomen would not dream of using convenience foods. Incoming women, in particular, defined 'real' rural women as possessing a rural background and being able to cook, bake and generally be domestic. Women living in the villages prescribed very traditional domestic identities when they were describing what they perceived to be true countrywomen.

These women drew clear links between domesticity, rurality and womanhood.

When one is drawing out the links between rurality and domesticity it is interesting to note that some incoming women stated that the reason for moving into the village was to be more 'homely'. These women drew direct links between domesticity and rural lifestyles. In the final section of this chapter these links are explored through a discussion of the attitudes and experiences of some incoming women.

IMAGINING RURALITY: INCOMERS AND THE 'DOMESTIC IDYLL'

Although local women were beginning to 'get with the times', as one woman suggested, and undertake formal employment, some incomers argued that the prime reason for moving to the village was to get away from the pressures of modern living. A minority of people moved into Ditton and Llangeley to live what they believed to be 'authentic rural lifestyles'. It became evident that part of this 'authenticity' entailed getting back to nature and 'the domestic'. This is not to suggest that this is the only reason people moved to rural areas nor that they lived out a domestic utopia when they arrived. Lizzy moved out to Llangeley with her husband thirteen years ago from the south of England. Their children had grown up and they were both self-employed, so they decided to live what she termed a more 'peaceful and quiet domestic existence'. Lizzy said that she and her husband were pretty self-sufficient in terms of growing their own food, and until recently they had had a cow whose milk they used to make their own butter and cheese. Lizzy continued:

'I bake my own bread ... I enjoy baking and I do those sorts of cookery things. I make my own puff pastry and jams and pickles. . . . I go shopping for what I call real things. I don't usually get a lot of frozen pre-cooked things, we used to make our own sausages and things like that. . . . Still, sometimes I am envious of people that pop into Iceland ... and stock up their freezers ... whatever I cook I start from scratch *but that is what we came for*' [emphasis added].

Lizzy makes the point that more and more people are moving to the countryside to live what they perceive to be a country life. Elaine also noted the small, but growing, minority of what she terms 'hippie types'. She believes that part and parcel of this dream was to get back to nature and become more domestic. For example, Pauline, a mother who had moved into the area two and a half years ago with her husband, wanted to get away from the pressures of city life: 'Before I moved I had a high-pressure job and I felt that I wasn't spending enough time with the family ... we thought this area could offer us a change, to get away from

133

all that.' Pauline felt that village life offered her space and an opportunity to spend time with her family. Although she categorically stated that she had no intention of spending all her days doing housework she wanted to be more 'homely'. However, Pauline's husband did not give up his job, like she did, but rather commuted long distances three times a week. Pauline explained that being at home with her children was part and parcel of her feelings of getting back to the 'natural', and although she agreed that she could have given up work and stayed where they were, 'it wouldn't be the same'. Indeed, she felt that there would be more pressures on her from working friends to go back to work if she had continued living in an urban environment. This may reflect a trend acknowledged by Little (1994: 26) in which 'Popular perceptions about rural living appear to create a greater acceptance or tolerance of the lack of employment opportunities for women.'

Lizzy argued that women who are more 'urban-minded' would not dream of moving to an area like Llangeley. She continued, 'You have to want to live in these types of circumstances. If you are urban-minded and want to fill your freezer with ready-cooked meals, it is quite different.' Lizzy seems to be suggesting that urban-minded women are less domesticated while on the other hand 'rural-minded' people prefer a more domestic lifestyle. Jenny argues that rural life attracts a certain kind of woman, having little to offer a woman without a family:

'I mean, if I was young like you there is no way that I would live in the country 'cause you would be so restricted. . . . You just wouldn't think about it, would you? Rural life has nothing to offer young people . . . but I think once you were married and had a family then you would think about it . . . and fit in.'

Indeed, during the research no single-career women were interviewed or even discussed by other women. In fact, all the women who had moved into the village were married with children or else retired. Although the career opportunities were limited in the area, the lack of career women in the village seems to reflect more general attitudes concerning what rural life offers. Indeed, my own experiences during the research reflected how women in the village viewed 'women alone'. Their concern and apprehension for me as a woman living alone, however touching, was very revealing in itself. According to Jenny, only when I was married and had a family would I really 'fit in'. The minority of women who stated that their main reason for moving was to live a more domestic existence were all financially secure. They were also married with children (or grown-up children). Although, as Elaine agreed, the numbers of one-parent families are growing in rural areas, as elsewhere, the attitudes of women in the study and the prevalence of traditional constructions of femininity beg questions about the acceptance of women on their own, either single parents or career women.

CONCLUSION

Drawing on recent literatures in both rural and feminist geographies concerning the socially constructed, and historically specific, nature of rurality and womanhood, this chapter has endeavoured to highlight the complex inter-linkages of domesticity, femininity and rurality as they interrelate to influence the domestic identities of contemporary rural women. By focusing on the domestic realm this research has recovered the historical and contemporary lived experiences of rural women, experiences all too often neglected and marginalised within rural writings.

It is clear that, traditionally, women in rural areas have, indeed, played an important role within the reproductive sphere. This was reflected in discussions with women who had lived in Ditton and Llangeley all their lives. Domesticity was central to their gender identities and their feelings of self-worth. However, it is also clear that the prevalence of traditional gender identities had influenced younger women, like Emma, and also the attitudes of women moving to the countryside more generally. Dominant social codes have constructed the home as a woman's 'natural' place, and this idea was sustained through the traditional ideas of women (and men) living in the countryside. Because Ditton and Llangeley are very small villages, women who did not conform to the moral code were not only noticed, but also judged. This raises questions concerning the possible marginalisation of women who do not 'fit' traditional feminine identity patterns in contemporary rural society.

Incoming women drew clear links between domesticity and rurality both in their discussions of whom they would define as 'rural women' and also through their own ideas concerning 'authentic' rural living. Moreover, this constructed norm was reinforced by the images possessed by incoming women concerning rural living.

However, the linkages between domesticity, femininity and rurality were not simple. With social divisions such as age and class weaving through these relationships the situation becomes increasingly complex. Women's domestic identities are shaped not only by the (dominant) moral order and social codes but also through economic necessity. Some rural women had to go out to work, thereby challenging the traditional orthodoxy. This research has highlighted the fact that this is not a widespread contemporary phenomenon although it has become increasingly common as farm incomes have decreased and, as many women in the study have argued, rural society becomes more materialistic. However, women's paid employment tended to be part-time and fitted around domestic responsibilities rather than an outright challenge to women's domestic traditional domestic identity. This, to some extent, is personal choice and must be viewed as women contesting and challenging their perceived place in the home. However, it is argued that this choice is restricted and constrained by expected norms and behaviours in rural life. It is women like Pauline who are best able to exploit the symbolic and substantive linkages between domesticity, rurality and womanhood; she had the financial means to move to a village in the first place and, secondly, to stay at home with her children once she had arrived.

If we are to accept that gender identities are unstable, historically specific and constituted in place, we must be wary of extrapolating these findings to 'other' rural places. Ditton and Llangeley were located in a strong agricultural area which possessed particular sets of gender relations surrounding the orthodoxy of family farming. The historical and geographical situatedness of socio-cultural constructions of rural womanhood must be acknowledged and investigated if rural experiences are to be successfully understood. Clearly, much more research needs to be undertaken on the importance of gender relations and the experiences of gender relations within rural contexts. However, future research, while taking account of gender divisions, must also be sensitive to the diversity of female (and male) experiences in contemporary rural society.

NOTES

1 Feminist theories and feminist geographers have converged in their interest in spatiality and place identity in attempting to think through, and build on, the diversity of female experiences.

2 The names of the villages and the names of all the interviewees have been changed for the purposes of anonymity.

3 Such epistemologies critique the principles of Enlightenment scholarship which attempted to lay bare universalised truths revealing the basic features of social reality. Feminists challenge this notion, arguing that 'what had most frequently been presented as objective because supposedly devoid of the influence of values, such as those related to gender, actually had reflected such values' (Nicholson 1990: 3).

4 Having discovered that there was very little academic literature concerned with rural women's lives, I rejected the use of a survey method as this would necessitate adopting pre-coded questions which, I felt would define women's experiences at the outset.

5 As a result of the fact that the locality has been dominated by agriculture, the majority of women who had lived in Ditton and Llangeley for most of their lives were, or had been, involved with the agricultural community, being either farmer's wives or daughters of farmers.

6 This reflects national trends of women's employment in rural areas (Little *et al.* 1991).

REFERENCES

Braithwaite, M. (1994) *The Economic Role and Situation of Women in Rural Areas*, Luxembourg: Office for Official Publications of the European Communities.

Brandth, B. (1994) 'Changing femininity: the social construction of women farmers in Norway', *Sociologia Ruralis* 34, 2: 127–49.

Bunce, M. (1994) *The Countryside Idyll: Anglo-American Images of Landscape*, London: Routledge.

Cloke, P. and Milbourne, P. (1992), 'Deprivation and lifestyles in rural Wales – II Rurality and the cultural dimension', *Journal of Rural Studies* 8, 4: 359–71.

Crouch, D. (1992) 'Popular culture and what we make of the rural with a case study of village allotments', *Journal of Rural Studies*; 8, 3: 229–40.

Davidoff, L., L'Esperance, J. and Newby, H. (1976) 'Landscapes with figures: home and community in English society', in Mitchell, J. and Oakley, A. (eds) *The Rights and Wrongs of Women*, Harmondsworth: Penguin.

Glenn, E. N. (1994) 'Social constructions of mothering: a thematic overview', in Glenn, E. N., Chang, G. and Forcey, L. R. (eds) *Mothering: Ideology, Experience and Agency*, New York: Routledge.

Halfacree, K. (1993) 'Locality and representation: space, discourse and alternative definitions of the rural', *Journal of Rural Studies* 9, 1: 23–37.

Hall, C. (1992) *White, Male and Middle Class: Explorations in Feminism and History*, Cambridge: Polity Press.

Little, J. (1986) 'Feminist perspectives in rural geography: an introduction', *Journal of Rural Studies* 2, 1: 1–8.

—— (1987) 'Gender relations in rural areas: the importance of women's domestic role', *Journal of Rural Studies* 3, 4: 335–42.

—— (1994) 'Gender relations and the rural labour process', in Whatmore, S., Marsden, T. and Lowe, P. (eds) *Gender and Rurality*, London: David Fulton.

Little, J., Ross, K. and Collins, I. (1991) *Women and Employment in Rural Areas*, London: Rural Development Commission.

Moore, H. (1988) *Feminism and Anthropology*, Cambridge: Polity Press.

Mormont, M. (1990) 'Who is rural? How to be rural: towards a sociology of the rural', in Marsden, T., Lowe, P. and Whatmore, S. (eds) *Rural Restructuring: Global Processes and Their Responses*, London: David Fulton.

Nead, L. (1988) *Myths of Sexuality: Representations of Women in Victorian Britain*, Oxford: Blackwell.

Nicholson, L. (ed.) (1990) *Feminism/Postmodernism*, New York: Routledge.

Philo, C. (1992) 'Neglected rural geographies', *Journal of Rural Studies* 8, 3: 193–207.

Redclift, N. and Whatmore, S. (1990) 'Household, consumption and livelihood: ideologies and issues in rural research', in Marsden, T., Lowe, P. and Whatmore, S. (eds) *Rural Restructuring: Global Processes and Their Responses*, London: David Fulton.

Rose, G. (1993) *Feminism and Geography: The Limits of Geographical Knowledge*, Oxford: Blackwell.

Stebbing, S. (1984) 'Women's roles and rural society', in Bradley, T. and Lowe, P. (eds) *Locality and Rurality*, Norwich: Geobooks.

Stone, M. (1990) *Rural Child Care*, London: Rural Development Commission.

Whatmore, S. (1990) *Farming Women: Gender, Work and Family Enterprise*, London: Macmillan.

Whatmore, S., Marsden, T. and Lowe, P. (eds) (1994) *Gender and Rurality*, London: David Fulton.

8

EMPLOYMENT MARGINALITY AND WOMEN'S SELF-IDENTITY

Jo Little

INTRODUCTION

Despite a lack of widespread or particularly detailed data, research has succeeded in drawing attention to the scarcity of job opportunities for women within the rural labour market (see, for example, Rural Development Commission 1991; Little 1994; Townsend 1991). Academics and policy-makers alike have construed this scarcity as 'a problem' given a societal context in which women are making an increasingly important contribution to the labour market. The lack of employment opportunities is seen as particularly problematic for certain groups in the rural community; for example, working-class women, single mothers and young women. Where paid work does exist for women locally, it is argued, that work tends to be poorly paid, unskilled, part-time or temporary. Moreover, the ability of women to take up those opportunities that do exist is frequently hampered by other structural problems such as low levels of public transport and childcare facilities.

Outside this fairly general recognition of employment opportunity, our knowledge and understanding of women's experiences within the rural labour market remains limited and poorly theorised. There have been few attempts to look beyond the basic characteristics of women's participation and non-participation in paid work in any broader examination of, for example, career histories and aspirations, job satisfaction, etc. More striking, however, has been the failure to consider the role of women's employment in terms of the construction of social relations and gender identities and the negotiation of household strategies in rural areas. We therefore know little about how women's paid work intersects with their domestic and community roles (other than in very general terms of time management), nor how this work is interpreted within any broader operation of rural society. It is important that our consideration of women's work (paid and unpaid) now starts to look at the totality of their lives and, in so doing, to appreciate the impact of people's expectations, assumptions and understandings of the rural community on women's involvement in the labour market.

The intention of this chapter is, therefore, to examine the notion of marginality as it relates to women's employment involvement, not simply in terms of job availability and conditions but also in the broader context of rural lifestyles and gender identities. It sets out to identify wider constraints on women's participation in paid work including the negotiation of gender roles as part of household strategies and community organisation. Throughout the chapter, attempts will be made to show how the distribution of economic roles within the rural household and community and the circumstances and experience of labour market participation for women are fundamentally bound up in the myths, assumptions and expectations which surround rural society and culture. People's understanding and rationalisation of employment is both a function and a part of their way of making sense of and living the rural. Women may be disadvantaged, it is argued, by the absence of particular jobs and by the actual characteristics of the labour market in rural areas.

The marginalisation of their employed identities is further cemented by the dominance of a popular rural culture which elevates the 'natural' women's roles and skills within the family and community, and separates and alienates their identities in the 'public' sphere of waged work. Just how this culture is reproduced within contemporary rural society is one of the main questions this chapter seeks to address.

THEORETICAL APPROACHES TO THE STUDY OF RURAL WOMEN'S EMPLOYMENT

The following section outlines some key theoretical and conceptual issues, mainly from feminist perspectives, that have influenced the research and writing of this chapter. The section does not attempt to provide a comprehensive review of different feminist approaches but rather seeks to show how certain arguments and issues of relevance within contemporary feminist geography have informed and shaped the direction and conclusions of this research. It goes on to demonstrate how these contemporary feminist approaches relate to other areas of current theoretical interest within rural geography, thereby tying this research more directly to the themes developed throughout the book.

Recent work by Hanson and Pratt (1992, 1995) (see also Pratt and Hanson 1990) has looked in detail at the relationship between gender and space as played out through women's labour market experiences. In their work they are concerned with questions of gender identity and the ways in which 'local places and identity intersect, overlap and shape each other' (1995: 22) and how this in turn influences and is influenced by patterns of employment participation. While Hanson and Pratt's research focuses on an urban area, the town of Worcester, Massachusetts, in the United States, there are many parallels to be drawn between their study and the present research, and much of the theory which underpins their arguments is equally relevant to the rural context of this chapter.

In establishing the theoretical framework for their empirical work, Hanson and Pratt stress the need to go beyond what now may be seen as 'traditional' studies of geography and women's employment, which have tended to focus on issues of distance and separation. Such studies have seen the importance of place primarily in terms of the spatial relationship between home and work and the isolation of the private domestic world from the public sphere of waged employment. Exceptions from the 'localities' tradition have emphasised the variations that exist in the gender division of labour from one region to another – the ways in which, for example, different places exhibit different patterns of employment segregation as a result, at least in part, of varying historical and political traditions, cultures and social relations (see, for example, Massey 1994). But even these studies tend to see space more as a container for gender and other social relations than as something fundamental to their formation and development.

What Hanson and Pratt suggest is that recent feminist concerns with subjectivity and the body and its relationship with space can lead to a much richer reading of the construction of different gender identities within the urban environment which in turn greatly enhance our understanding of women's experiences within the local labour market. They cite work (for example by Pollock 1988) in which the spatial mobility of women is interpreted in the context of feminine bodily comportment and the 'gaze' of men in the city. They go on to argue that as well as recognising the importance of the relationship between the body and the urban form, we must also acknowledge how such a relationship has been 'inscribed and rescribed through discourses, cultural representations and everyday practices' (Hanson and Pratt 1995: 18).

Hanson and Pratt put forward what they describe as a model of the 'reciprocal constitution of subjectivity and geography' in which, rather than seeing the urban form as simply constraining and limiting the spatial territory of women's bodies, a greater interaction between bodies and cities has evolved in which, for example, our own beliefs about our bodies serve to limit our use of space and our expectations of spatial boundaries and constraints. Such a model has important implications for the way in which we conceptualise women's involvement in the local labour market. It

> prompts us to think more fully about the creation of different gendered and social identities in different places and to look at and listen carefully for the many different ways in which these identities are inscribed. It pushes us beyond considering geography and local

labour markets only in terms of opportunities and constraints, as envelopes of resources that allow or disallow individuals to fulfil their preconceived potentials; it opens the recognition that gendered identities, including aspirations and desires, are fully embedded in – and indeed inconceivable apart from – place and that different gender identities are shaped through different places.

(Hanson and Pratt 1995: 18)

The focus in feminist geography on the relationship between subjectivity and space and on gender identity is compatible with recent theoretical discussions within rural geography and social studies. As noted elsewhere in this book and in the literature (see, for example, Cloke and Milbourne 1992; Mormont 1990), rural research has begun to look much more seriously at the relationship between social relations and space and at the specificity of a *rural* dimension to associations between place and identity. Previous disenchantment with 'the rural' has given way, in some areas, to an appreciation of the importance of distinctly rural ideas and images to cultural construction of place and to their representation (see, for example, Bunce 1994; Crouch 1992). These various debates in feminist and rural studies together provide a useful framework for the examination of gender identities in rural areas and how such identities relate to rural spaces generally and to specific rural places.

In summary, acknowledging that understanding women's position within and experiences of the local labour market involves looking beyond the implications of distance and the separation of home and work takes us into wider questions surrounding subjectivity and the gendered use of local spaces. How the relationship between gender identity and space is played out is critical, it is suggested here, to the choices available to women within the local labour market and how they interpret and experience those choices. Any attempt to explore in depth the basis of the relationship between space and gender identity demands sensitivity to the meanings of particular places to particular people. In a rural context, then, the cultural constructions of rurality and the association between images and ideas of the rural and the everyday lives of women and men contribute to (and, in turn, are a product of) employment experiences, expectations and histories.

Finally, it is important to recognise here the progress that has been made in addressing the concept of difference and moving away from the idea of a single gender identity shared by all women. A growing body of theoretical and empirical research by feminist social scientists, much of it inspired by the postmodern 'turn', has argued the value of recognising the multiple and shifting identities of women (see, for example, McDowell 1993; Bondi 1994; Nicholson 1990; Pratt and Hanson 1991). Such work has begun to acknowledge the different experiences of women of different classes, races, sexualities, etc. and served, as Gibson-Graham (1994: 213)

141

notes, to 'dissolv[e] the presumed unity of women's identity'. While a certain amount of panic has been created among feminists who have seen the legitimacy of past research and political practice as resting on the assumption of a common subordinated identity, deconstruction has also given rise to fruitful theoretical debate as to who women are and how we know them (Gibson-Graham 1994).

The intention here is not to embark on a lengthy critique of the various views or positions on 'difference' as either experienced by women or theorised by the academy. Rather, the purpose of these comments is to acknowledge the importance of recent developments in feminist theory to the chapter's concern with gender identity. Thus, underlying the observations and assertions made here is a sympathy with the notion of multiple identity and a firm belief that women do not share a single gender identity. Having made this point, however, the chapter also rests on the contention that certain aspects of gender identity may be shared by women in particular places at particular times. A recognition of difference – 'the surrendering [as Gibson-Graham 1994: 213 puts it] of epistemological claims about women's shared identity' – should not mean that we reject any possibility of commonality between women or that we throw out the feminist project which is to challenge the relative powerlessness of women as women.

In the context of this chapter, it will be argued that there *are* certain characteristics that are shared, or at least (importantly) are perceived to be shared, by women in rural areas. Indeed, it is just these expectations of a particular and common identity which constitute such a powerful influence on gender relations and hence on women's behaviour, use of space and employment participation.

This brief theoretical discussion has opened up a number of different areas of debate for exploration in this chapter. Clearly, some of the issues raised are complex and far-reaching, and cannot be concluded here. It is important, however, that studies of rural women and marginality do begin to take on board issues of identity and that in doing so they draw on the wealth of feminist research which considers how identities are negotiated and sustained. This chapter provides an initial attempt to interrogate some of the major facets of *rural* women's identities and to link an understanding of identity to employment participation. In attempting to do this it argues, both explicitly and implicitly, that the importance of space and place must be acknowledged. At a general level the chapter attributes a powerful formative influence to rurality itself, arguing that women's experience and understanding of rurality is crucial to their identity. At a micro-level the interaction between place and particular spaces within the village and gender is also important. In the discussion of Hanson and Pratt's work above, attention was directed to the relationship between space and the body – using the notion of bodily containment. Here 'the body' is taken at a somewhat broader level in the exploration of women's use of and involvement within certain 'spaces' of the home and village community.

WOMEN AND THE RURAL LABOUR MARKET: EVIDENCE FROM THE RURAL COMMUNITY

The chapter will now turn, via the analysis of material collected from two 'case study' villages in Avon, to a consideration of rural women's employment and lifestyles. The case study discussion will not only identify women's current position within the rural labour market but also attempt to locate women's employment within a much broader analysis of their gender identities within the villages as introduced above. Aspects of their roles within the community, including their participation in voluntary work and in village 'societies' and 'events', will be considered as well as their attitudes towards rurality and attachment to the village.

The two villages where the research takes place are both located in the 'accessible' countryside around Bristol. East Harptree lies eight miles to the south of the city in the Chew Valley – an Area of Outstanding Natural Beauty, highly sought after by commuters and other incomers. Hawkesbury Upton is a similar distance from Bristol to the north-east of the city. Again, its location and attractiveness make the village popular with commuters. Hawkesbury Upton is the larger of the two villages (with a population of over 800 to East Harptree's 633) and tends to be the more dynamic, having had more recent housing development within the village boundaries. Both villages have a number of basic services including a village shop and post office, a primary school, village hall, pub and some sort of bus service. Neither, however, has much to offer in terms of formal employment opportunities; the only local employers (other than occasional jobs in the pubs, shops or school) are an egg farm and a yoghurt factory in a few miles from East Harptree and a retirement home and hotel in Hawkesbury Upton.

The data used in this chapter were obtained through semi-structured interviews and questionnaire survey, undertaken in 1992 and 1994.[1] Basic information on employment participation, career histories as well as household size and structure was recorded in a questionnaire survey carried out among women of employment age in both villages. In total, eighty-five questionnaires were completed. Additional information, much of it attitudinal, some 'anecdotal' but all very relevant and important, was gathered through longer interviews with selected women (and the occasional man!). These (twenty) interviews were semi-structured in that they covered a range of pre-determined areas of interest but each took its own direction depending on the views and experiences of the particular interviewee. The replies cannot be described as 'representative' of all women or indeed of any specific group of women; the information provides insights into social and economic conditions and characteristics in the villages and into attitudes, values, beliefs and experiences which could not be gained through a more structured approach. While care must be taken in the interpretation of such material, the richness it provides in dealing with the qualitative issues at the heart of this research cannot be denied.

Women's employment participation

The basic detail of women's participation in the rural labour market, as identified by this research, reveals few surprises, generally conforming to the findings of previous (albeit numerically limited) studies. Thus the Avon research found fewer women in the sample to be engaged in paid work than is the case nationally, with an overwhelming majority employed part-time. Sixty per cent of women interviewed had a paid job (compared to a national figure in 1991 of 71 per cent of all women) and 75 per cent of these were part-time (as opposed to 42 per cent nationally). In East Harptree and Hawkesbury Upton there were higher numbers of women employed in professional jobs than is generally the case in rural areas (see Table 8.1), and, partly as a result, conditions of work were found to be 'better' than has been recorded elsewhere (although the problem of measurement means that such claims must be treated with caution). The relatively high proportions of women working in teaching and the health service meant, for example, that amounts of paid holiday, levels of sickness pay and the proportion of jobs carrying formal contracts were significantly higher than has been found elsewhere (see RDC 1991).

Table 8.1 Employment of respondents by type

Employment type	% of respondents
Managerial	3
Professional (e.g. teaching/medical)	39
Clerical/secretarial/nursing	39
Semi-skilled	11
Unskilled	8

Source: Questionnaire survey, 1992–4

A characteristic of women's employment participation in East Harptree and Hawkesbury Upton that did agree with findings from other rural areas was the local concentration of their jobs. Table 8.2 compares the location of employment of male and female villagers from East Harptree only (to avoid over-complication). It demonstrates the significance of the surrounding villages for women's employment and of major urban locations (somewhat further away from home) for men's. Interesting is the fact that despite the more local nature of women's participation generally, more men

Table 8.2 Employment location of women and men (East Harptree only)

Location of workplace	Women (%)	Men (%)
From home	5	18
Chew Valley	49	18
Bristol, Bath, Weston-Super-Mare	29	48
Other local towns	12	9
Other	5	5

than women worked from home. These men were mainly professionals: self-employed solicitors, accountants and consultants. Except in the case of a builder and a couple of farmers, they did not run 'family businesses' but worked alone, employing no one (although often making use of help from their wives for typing and book-keeping).

One feature of women's employment participation in rural areas that has received relatively limited (and very superficial) attention is the extent to which women who are employed work in jobs that do not reflect their training and qualifications. This is clearly linked to levels of job opportunity, but, while closely related to the availability of paid work, cannot be fully explained by it. It incorporates important and complex relationships between the household and the community and the labour market, and can provide a very interesting way into discussions about choices and constraints in respect to women's employment. This gap between employment experiences and qualifications is also critical, it may be argued, in the formation and sustaining of gender identities and, in the context of broader employment histories, highly relevant to the negotiation of household strategies and lifestyle decisions.

Almost half the women interviewed said that they had professional or academic qualifications equivalent to A levels and beyond (19 per cent had a degree). A further 28 per cent had O levels or the equivalent. Fifty-two per cent of those currently in a paid job were doing work which was not that for which they had been trained or did not use the qualifications they had achieved. While it is recognised that many people, for a wide variety of reasons, do not end up following the career for which they initially train and that this is not necessarily evidence of an absence of choice, the high proportion of women working in jobs which required a lower level of formal qualification or training than they possessed can be seen as particularly noteworthy in respect to the debates raised here concerning gender roles and identities within a rural context.

A part-time doctor's receptionist post had been advertised shortly before we conducted the research in East Harptree. There had been 'floods' of well-qualified applicants – women who wanted to work very locally and only during school hours. The post was not very well-paid or 'fulfilling' according to the practice manager, but it did provide the opportunity to work in a 'nice' environment just outside the village. As one respondent summed it up: 'The main difficulty is finding employment that will fit in with my children's school hours. Jobs that fit that description are very sought after and few – thus they are not easy to get.'

When questioned directly, the women interviewed mostly claimed to be satisfied with their current employment situation (although 23 per cent did say that they would change their jobs if something suitable came up). Many did, however, cite problems that they felt affected 'other groups' in the rural community. The apparent or declared satisfaction masked the fact that employment choice was highly restricted in terms of both the type or variety of work available and the seniority and status of the jobs that did exist. For professional women one of the main options was teaching (with 20 per cent of the employed respondents currently working as teachers). The advantages of working hours that were guaranteed, almost by definition, to fit in with their children's school day and term times was mentioned as the main attraction of teaching (both by those in the profession and by those outside). Not all respondents working as teachers had been able to get regular, full-time jobs where or when they wanted to, and several worked as supply teachers and consequently experienced considerable uncertainty in terms of their working week and their future career progression. Some women expressed a wish to leave teaching (citing the lack of support and regard for teachers and the poor financial rewards) but recognised the difficulties of finding another professional job in the area that would fit in with the school day. Outside teaching, women in professional jobs generally commuted to Bristol or to other urban centres (only three women worked 'locally' – in the village or surrounding area).

Despite the clear difficulties experienced by professional, often older, women in gaining appropriate employment, it was among the younger women, especially the school leavers, that the greatest difficulties were generally perceived to exist. The importance of this perception itself in the context of gender identities in the rural community is discussed below. What is important to note here is the regularity with which the absence of jobs for young people in the villages and surrounding area was recorded. The main concern in this context was that the jobs which existed in the immediate locality and that were at a junior level were almost exclusively 'unskilled' and poorly paid. The only options therefore for young women from East Harptree leaving school without A levels or employment-related qualifications were in the yoghurt factory, mushroom farm and egg-packing factory, while in Hawkesbury Upton the choice appeared to be between the local hotel and pub or a residential home for the elderly. For those looking for 'career-type' jobs (that would offer training and possibly qualifications) and those wanting to return to the area after college or university, prospects were extremely limited.

Although it was extremely difficult to get reliable information, all the women interviewed in Hawkesbury Upton referred to the existence of a shadow or informal economy in their village. Working-class, generally younger, women, it seemed, were frequently involved in child minding and catalogue sales as a means of making money. For such women the expense of childcare for their own children as well as the scarcity of local employment forced them to rely on *ad hoc* and casual sources of income. It was widely acknowledged that without this income many of the women would not be able to live in the village in their present circumstances.

Family and household

As discussed above, feminist theory has stressed the importance of exploring the spaces of the community and the home in the explanation of labour market segmentation. The choices available to and made by women in entering the labour market incorporate and reflect the social and gender relations of the household and the community as well as those of the economy and workplace. Looking at the intersection of labour market, household and community rests on a detailed understanding of the broader lifestyle and life cycle strategies negotiated within the household, the assumptions and expectations of gender relations in the community and the ways in which the operation of both the household and the community construct, reflect and reproduce gender identities. In the second part of the case study these issues will be discussed as they contribute to the examination and comprehension of women's experiences of the rural labour markets of East Harptree and Hawkesbury Upton.

The first task here is to look beyond women's immediate labour market involvement at their longer career histories and to place these within the broader evolution and decision-making of the household. In this way we can examine the context within which decisions about employment are made and start to identify the range of pressures and preferences which surround the changing commitment of women to the labour market.

The employment histories and career paths of the women interviewed in the course of this research were highly fragmented – especially among those with children (85 per cent of all respondents). Seventy-six per cent of women with children had had a paid job before having children but less than one-third had gone back to the *same* job when they went back into the labour market. Those who did take maternity leave and return to the same employer after the birth of their child tended to fall into one of two groups. Either they were professional women in senior (generally managerial) positions – for example, a bank manager and a GP – or they were women who worked as nannies or in small family businesses where they could take their children to work with them.

Clearly, there is little peculiar to a rural lifestyle in the points made here (nor in the case histories cited). The only obviously (but not exclusively) rural influence was an absence of accessible child-minding facilities, which was mentioned by a small number of women as a factor in their decision not to return to the labour market after the birth of their child or children. Seen in a broader context, however, decisions about the re-entry into employment by women were part of household lifestyle strategies in which certain assumptions and perceptions of the 'rural' were an important feature. The spaces of the rural household and community could indeed be seen to exert an influence on women's employment opportunities and decisions.

Since the majority of respondents were 'incomers', it was felt to be important to consider how and where decisions to move to the village fitted in with women's employment histories/stories. Interestingly, while 23 per cent of respondents cited 'husband's job' as a reason for moving to the village, none mentioned their own employment as playing a part in decisions to move. There are often complicated circumstances around families' relocation strategies and it is important that motives are not interpreted too simplistically. What emerged from the research, however, was that for many families, moving to the village incorporated a set of lifestyle decisions and the adoption of a series of priorities which were made possible or enhanced by the opportunity to live in a rural community. So, in articulating reasons for moving to the village a bundle of household, lifestyle and rural attributes were mentioned. For some respondents it proved, not surprisingly, impossible to separate out these interrelated factors.

The relocation of the husband's job (or another change such as promotion or the capacity to work from home) and the subsequent move to the village had, in a number of cases, coincided with women stopping work and having children. The desire to give up paid work and to stay at home while their children were small (until, in effect, they started school) was expressed by many mothers. What was particularly interesting here was the extent to which this desire embraced particular responses to the rural environment and community. Certainly it reflected some recognition of the practical difficulties of obtaining local employment and juggling responsibilities at home and at work in the countryside. But, more significantly, it appeared to reflect many of the assumptions of the rural idyll and of the particular role of women as wives and mothers in the rural family. Annie Hughes' chapter (Chapter 7 in this volume) explores in more detail the importance of women's domestic role as a part of the rural idyll (see also Little and Austin 1996).

As 'Jenny's' case history illustrates (see boxed text), rural areas were undoubtedly seen as very favourable environments in which to bring up children in that they are perceived as places in which 'family values' have endured. The importance of the mother's role and of the tight-knit family unit and their survival within *rural* social relations were frequently cited

One woman, 'Jenny', from Hawkesbury Upton, described how she and her husband had moved into the village before the birth of their first child. Jenny had given up her paid job then and had not returned to employment since (their eldest child was now at school and they had two others). The couple had been keen to move to the country-side to allow their children to experience the 'advantages' of growing up in a village, and, despite the fact that Jenny's husband now commuted considerable distance to work and was frequently away overnight, their expectations of village life had been fulfilled. Jenny felt that her children had more freedom and were involved more in community life than they would have been in a town. She also saw it as essential that she be at home with the children to enable them to participate in village activities, play with village friends, etc. While Jenny was anxious to return to full-time employment, she felt that doing so would not be 'fair' to her children (especially the youngest).

as part of the attractiveness of living in the countryside. Two women in Hawkesbury Upton, for example, spoke proudly of the fact that 'there are no latchkey children in this village'. It is inappropriate to talk in quant-itative terms about the extent and pervasiveness of such views or of the relationship between a belief in certain 'qualities' of rural living and prac-tical household strategies and decision-making. It is important, however, that something of the strength, ubiquity and comprehensiveness of these associations be appreciated and acknowledged as a key influence in women's labour market involvement and choices.

The power of the rural ideology in elevating and defining the role of mothers and its impact on women's suitability for employment is not simply an issue in terms of *self*-identity. The appropriateness of rural women's involvement in the labour market is, respondents claimed, reflected in the attitudes of existing and potential employers. Several women felt that assumptions concerning the priorities they attached to family life and caring for their children had affected their treatment by employers. See the case of 'Liz' below.

Not all women interviewed were in a position to chose whether or not to take on paid work; 8 per cent of employed women were the sole income earners in the household (only two of these lived alone), and for them, staying at home with their children all day was not an option. Some made particular arrangements at work to enable them to be at home when their children came home from school (such as working night shifts at the local factories), and many relied heavily on relatives to collect and care for their children. While considerable sympathy was expressed over the plight of

'Liz' lives in Hawkesbury Upton. She is a mother of three children aged between 3 and 8. She is self-employed, working as a freelance writer of computer manuals. She does her paid work mainly while her children are at school and in evenings and at weekends. She also makes complicated (and reciprocal) arrangements with other mothers over collecting children from school and after-school care. Being at home on at least some afternoons to collect her children and others from school is important to her and not something she would want to give up. Liz spoke (rather angrily) about the way in which those who supply her with work assume that she will have difficulty completing work during school holidays. She feels that she is treated rather as a 'country bumpkin in a pinny – that "funny" woman living in a village' and while the quality of the work she does is never questioned, she feels that the combination of being a woman and living in a village means that she is not treated seriously. Recently she has stopped including the name of her house in her business address to try to create a 'less rural' image.

women on low incomes, the stigma attached to the use of child minders and of 'not being there' for children was expressed both implicitly and explicitly by village women during the course of the interviews.

The expectation of motherhood is a powerful part of the rural ideology and one which is obviously incorporated into household strategies for childcare and in women's experiences and ambitions in the labour market. Numerous accounts of women's careers taking second place to childcare responsibilities were given to us during the course of the research.

'Although it may not necessarily be a good thing, most of my [13-year-old] daughter's friends are considered old enough to look after themselves after school and in the holidays too. The strain with younger children is always on the mother to provide the childcare. Before my present job [as a secretary] I worked as a cleaner and drove a fish van to fit in with school hour' (respondent from Hawkesbury Upton).

Again, it cannot be argued that this is a specifically or uniquely rural characteristic. Links between employment choices and the rural lifestyle were, however, frequently made by the women themselves. There was reference, for example, to the assumed value of the traditional family as an important attribute of the rural way of life, as described above, and to the fact that only by staying at home could women allow their children to fully experience the benefits of such a way of life. A number of women, when asked about employment participation, recounted their lack of career progression and continuity with the comment that had they wanted to pursue a career, they would have stayed in the town – again making explicit a link between full-time childcare and rural living.

Women and the rural community

Another 'space' that must be considered in the creation of gender identity is the space of the rural community. It is argued here that, like the rural family, the rural community has a particular set of identifiable meanings and associations which impinge in a particular way on gender identity (and, hence, following the argument of the chapter as a whole, on women's economic roles and experiences). These meanings are also reinforced by a variety of practical characteristics which combine to ensure that the rural community constitutes a powerful force in the construction of gender identity and on women's participation in waged work.

An aspect of community of particular importance both to the formation of gender identity and the practicalities of women's labour market participation is voluntary work. Commitment to voluntary work may be a factor in debates surrounding the division of labour in the household and the community, and hence to the balancing and negotiation of roles by women. More significantly, perhaps, the assumptions that surround women's involvement in voluntary work are crucial in terms of the operation of the rural community and of women's roles within it. Elsewhere there has been some discussion of the nature, extent and relevance of voluntary work to the rural labour market and community (see, for example, Rogers 1987; Woollett 1990) and focusing in particular on the extent to which voluntary work can be seen to 'plug the gaps' left by the formal labour market and by declining service provision in rural communities. Not until more recently, however, has there been any more direct consideration of women and the voluntary sector or of the broader implications of voluntary activity to gender roles and relations in the rural community (see Little forthcoming; Moore-Milroy and Wismer 1994; Teather 1994). While we are starting to learn a little more about the extent of women's voluntary commitment, we are still uncertain about the way in which that commitment has contributed to (and indeed is a factor of) women's relationship with the rural community.

The Avon research found, perhaps not surprisingly, the involvement of women in voluntary work to be high: 36.5 per cent of women interviewed said that they took part in some form of regular unpaid work outside the home. Commitment varied considerably from less than an hour a week to as much as 16 hours (for a fuller discussion, see Little forthcoming). The sort of work performed as voluntary activity also varied from that typically

151

done by volunteers – i.e. collecting for charities or running church or village 'events' – to that which appeared to be a much more direct replacement for formal paid work such as assisting local primary school teachers in the classroom and providing support for mothers isolated at home with pre-school children.

What is pertinent here is, first, the way in which voluntary work was regarded very much as a female preserve (by both women and men residents) and, secondly, how deeply embedded the notion of voluntary activity appeared within cultural constructions of the rural community. To some extent the two issues cannot be separated in the sense that the way in which voluntary work has become constructed within the village is as a part of the female identity in the community. Respondents in East Harptree and Hawkesbury Upton made it clear that, almost by definition, voluntary work was undertaken by women, and that active participation in the community necessitated some degree of involvement in voluntary work.

Traditional perceptions of voluntary work were found to be prevalent among the rural women interviewed, and even though actual participation in the voluntary sector is typically concentrated in the middle classes, the perceptions of its place in the rural community were shared by middle- and working-class women. Ideas about 'helping others' and 'doing your bit' in the village were frequently voiced as central to the notion of the rural community. While recognition was made of the differing skills, family and labour market commitments which influenced women's ability to take part in voluntary work in the village, failure to 'muck in' at some level was keenly identified and generally frowned on (according to some respondents). This again was very gender dependent, for while the expectation was for women to help out in the community (with both more formal voluntary activity and *ad hoc* village 'events') whenever possible, men were given much more opportunity to 'opt out' (and hence were seen as particularly heroic when they joined in). Voluntary work offered women a route into the rural community – a way of becoming accepted which was especially important for those moving in with no existing ties in the village.

The relationship between women's voluntary work and their participation in the formal labour market in rural areas is complex. As with many other issues discussed in this chapter, it is overlain by class position. One middle-class respondent in Hawkesbury Upton told me that her voluntary work occupied most of her time and that she was 'too busy to have a paid job'. Other respondents talked of the huge commitment made by some village women to community work – and also to the pride and professionalism characterising such work. Those who got involved to this degree (those who were seen to 'run the village') were all wives of highly paid workers. They had no financial need for a paid job and consequently they

'Women hold the village together.'

'Without women you wouldn't have a village life – men aren't interested in that sort of thing. They don't mind the odd skittles evening but they don't really want to be involved.'

and, from one of the men in Hawkesbury Upton, 'Women are better at that sort of thing – it's more natural to them.'

had time and energy (and perhaps the need in terms of self-identity) to devote themselves to voluntary work.

Outside these extreme cases involvement in voluntary work was not, among the village women we interviewed, generally determined by commitment in the formal labour market. Participation in voluntary work was in fact higher among those with a paid job than it was among those who were not in paid employment; While 38 per cent of employed women undertook voluntary work, only 29 per cent of those without a paid job did so. It was suggested, however, that the availability of more part-time local employment would lead to a reduction in some sorts of community-based voluntary activity. This reduction would, it was stressed, be selective, with many of the essential tasks (such as helping at mothers and toddlers groups and organising key village festivals) continuing.

Women's gender identity as it related to the village community went beyond the issue of voluntary work to include a much broader interpretation of the spirit of the rural community. Women were seen, and saw themselves, for example, as in many ways responsible for the protection, preservation and continuation of the village community, both in general and in highly specific ways. The remarks quoted in the boxed text above are illustrative of the sorts of ideas that were expressed.

Numerous informal, *ad hoc* 'tasks' around the village which were clearly encapsulated within the notion of community spirit as articulated by respondents could be seen to fall to women. 'Keeping an eye on' certain residents – examples were mainly elderly and infirm people but included families where there had been recent illness or unemployment – was seen as part of women's role in the village. Similarly, galvanising support for special local events was largely gender specific (meaning women did it!). In Hawkesbury Upton, a small group of self-appointed women monitored the appearance of the village, putting pressure on residents who were felt to be 'letting the village down' in terms of the state of their gardens. While there was some difference of opinion as to whether such tasks were necessary (especially in the case of the latter), the fact that they were done was widely seen as advantageous to the 'spirit of community'.

Ensuring that the community spirit endured motivated women's involvement in certain activities and events which might not have interested them personally but which they saw as important to village society. Not unusually, a high proportion of women we spoke to said they belonged to the village Women's Institute (WI). Several of these claimed that they would not have considered joining a WI before they came to live in a village. They saw the WI organisation not only as a good way for them to get to know village people (especially when they first came to live in the village) but also as a means of bringing villagers together. It gave them a formal basis on which to activate village events and traditions. Although the WI, particularly in East Harptree, was criticised by a few as being 'middle-class' and dominated by 'incomers', there was general agreement over its importance in generating and sustaining community spirit. Some working-class women expressed some ambivalence about the sort of community spirit it generated but none went as far as openly to criticise the intentions of the WI.

SOME CONCLUDING REMARKS

These are just some of the ways in which, I suggest, women's identity has incorporated a particular construction of rurality. Quite evidently, it is never easy to articulate or even defend links between representation and practical outcome. Nevertheless, it is important that research endeavours to ground some of the more abstract theoretical ideas about the characteristics of rural lifestyles and values in practical realities of people's daily existences. The preceding sections have pointed to a variety of ways in which ideas about village social relations are translated into everyday shared practices which in turn impinge on, or even determine, women's involvement in paid work. Through tracing important beliefs about rural living and women's 'place' in the rural household and community the research has attempted to argue that links between gender identity and employment incorporate a specifically rural dimension.

In illustrating the ways in which accepted ideas about rural domestic and social relations are reflected in gender identity, the chapter has been at pains to stress that not all rural women share a common identity. As has already been pointed out, work by feminist geographers has helped to develop a sensitivity towards difference between women and raised awareness of the dangers of treating women as a single group. This chapter acknowledges the importance of such work and of the richer understanding that a sensitivity to difference can give us in researching gender identity. It recognises too the complexity of the notion of identity and the impossibility of fully comprehending either individual gender identities (in their entirety) or the process of identity construction.

What has been argued here, then, is not that all rural women share a common (and single) identity but that there are a number of characteristics that have assumed a powerful and central role in the expectation and assumptions of women's identity in rural communities. Put simply, rural women exhibit a range of different characteristics in terms of class, ethnicity, age, disability, etc. which have a key bearing on their identities as women. There are, however, a set of powerful assumptions which influence beliefs about gender identities in rural communities. There is an image of 'the rural woman' to which, clearly, not all women conform – and to which, perhaps, no woman conforms totally – but which nevertheless influences the behaviour, values and expectations of all rural women and men, and hence becomes incorporated in a very real sense within gender identities. This research has focused on two main areas where such an image, a common identity, has been influential. It has been argued that the expectation of women's roles as wives and mothers has featured very significantly within rural women's identities and that internal and external beliefs about the appropriateness of women's involvement in childcare and family reproduction has been prioritised over employment. Secondly, the idea of women as 'community-makers' has, perhaps in a more subtle way, assumed a central role in the identity of rural women. This idea has been reflected in a number of ways including the participation of women in voluntary work and the organisation by women of village events – again affecting their involvement in paid work.

Research identified instances where women were attempting to escape what they saw as the negative associations of the dominant image of rural women. The case of 'Liz' was cited as an example of where external expectations and assumptions were seen to positively disadvantage rural women in relation to employment. In other instances respondents talked about the assumptions that surrounded their automatic responsibility for childcare – particularly the care of young children. The assumption that women's employment would necessarily take second place to their child-caring roles was something that, as pointed out earlier, the vast majority of women we spoke to explicitly and implicitly supported. Many did assert that that assumption was more enduring and ubiquitous in a rural than in urban society, while some also recognised that (although attractive to them) the very fact that the assumed childcaring role of women was so strong potentially limited their choices as regards involvement in waged work or alternative rural lifestyles (particularly living in a rural community as a single, childless woman).

There were instances, it should be added, when what were seen as negative and choice-limiting external assumptions of gender identity by some were perceived by others as positive. At least three women separately

155

and spontaneously voiced the view that the image of rural motherhood as apparently held both internally and externally was one that they were very comfortable with because it removed from them the guilt they felt in *not* wanting a career or paid job. These women said that contemporary increases in women's involvement in the labour market, the publicity around full-time childcare, and so on made them feel as if they *should* want to put employment first. The expectations of gender role within the rural household and community vindicated their personal choices and made them feel less apologetic about their priorities. Clearly, this is another complex area, and one linked to what has already been argued in respect to the internalised images and assumptions around gender identity.

As part of the postmodern 'turn' in rural social scientific research attention has started to be directed to the lives and experiences of 'marginalised' groups in rural society – something that nearly every contributor to this volume has acknowledged. It has been suggested that among the assorted 'others' in the rural community are women and that as a marginalised group we must seek to examine their lives and their struggles against the rural norms of Mr Average (see Philo 1992). This chapter has argued for a more subtle and sensitive reading of women's marginality, arguing that marginality cannot be fully understood without some sort of examination of gender identity. Here the focus has been on employment – the chapter asserting that the identity assumed by and for the majority of rural women investigated is one in which employment is marginal. While the labour market reinforces both directly and indirectly women's disadvantage, it is the assumptions about the true identities of rural women which shape and sustain the priorities placed by women and the household on women's employment. This chapter has only just begun to explore how the different influences of gender identity (among women and men), household strategy and labour market participation interact. The conclusions that have emerged demonstrate the complexity of the issues and the potential richness of future investigation in terms of a broader understanding of social and economic relations within the rural community.

NOTE

1 The 1992 interviews were conducted with the help of Dr P. Austin, Department of Planning, University of Auckland.

REFERENCES

Bondi, L. (1994) 'Locating identity politics', in Keith, M. and Pile, S. (eds) *Place and the Politics of Identity*, Routledge, London.
Bunce, M. (1994) *The Countryside Ideal: Anglo-American Images of Landscape*, Routledge, London.

Cloke, P. and Milbourne, P. (1992) 'Deprivation and lifestyles in rural Wales', *Journal of Rural Studies* 8 (4): 359–71.

Crouch, D. (1992) 'Popular culture and what we make of the rural: a case study of village allotments', *Journal of Rural Studies* 8 (3): 229–40.

Gibson-Graham, J. K. (1994) ' "Stuffed if I know": reflections on postmodern feminist social research', *Gender, Place and Culture* 1 (2): 205–24.

Hanson, S. and Pratt, G. (1992) 'Dynamic dependencies: a geographic investigation of local labor markets', *Economic Geography* 68: 373–405.

—— (1995) *Gender, Work and Space*, Routledge, London.

Little, J. (1994) 'Gender relations and the rural labour process', in Whatmore, S., Marsden, T. and Lowe, P. (eds) *Gender and Rurality*, David Fulton, London.

—— (forthcoming 1997) 'Women and voluntary work in rural communities', *Gender, Place and Culture* 4 (2).

Little, J. and Austin, P. (1996) 'Women and the rural idyll', *Journal of Rural Studies* 12 (2): 101–11.

McDowell, L. (1993) 'Space, place and gender relations: Part II. Identity, difference, feminist geometries and geographies', *Progress in Human Geography* 17 (3): 305–18.

Massey, D. (1994) *Space, Place and Gender*, Polity Press Cambridge.

Moore-Milroy, B. and Wismer, S. (1994) 'Communities, work and public/private sphere models', *Gender, Place and Culture* 1 (1): 71–90.

Mormont, M. (1990) 'Who is rural? Or, how to be rural: towards a sociology of the rural', in Marsden, T. Lowe, P. and Whatmore, S. (eds) *Rural Restructuring: Global Processes and their Responses*, David Fulton, London.

Nicholson, L. (ed.) (1990) *Feminism/Postmodernism*, Routledge, New York.

Pratt, G. and Hanson, S. (1990) 'On the links between home and work: family strategies in a buoyant labour market', *International Journal of Urban and Regional Research* 14: 55–74.

Pratt, G. and Hanson, S. (1994) 'Geography and the construction of difference', *Gender, Place and Culture* 1 (1): 5–29.

Philo, C. (1992) 'Neglected rural geographies: a review', *Journal of Rural Studies* 8 (2): 193–207.

Pollock, G. (1988) *Vision and Difference: Femininity, Feminism and Histories of Art*, Routledge, New York.

Rogers, A. (1987) 'Voluntarism, self-help and rural community development: some current approaches', *Journal of Rural Studies* 3 (4): 353–60.

Rural Development Commission (1991) *Women and Employment in Rural Areas*, RDC, London.

Teather, E. (1994) 'Contesting rurality: country women's social and political networks', in Whatmore, S., Marsden, T. and Lowe, P. (eds) *Gender and Rurality*, David Fulton, London.

Townsend, A. (1991) 'New forms of employment in rural areas: a national perspective', in Champion, T. and Watkins, C. (eds) *People in the Countryside*, Paul Chapman Publishers, London.

Woollett, S. (1990) *Counting the Rural Cost*, National Council for Voluntary Organisations, London.

9

LITTLE FIGURES, BIG SHADOWS
Country childhood stories

Owain Jones

INTRODUCTION

In this chapter my aim is to begin to explore some of the implications of recent calls for the retrieval of rural children's geographies and in particular how such geographies are 'structured "from without" and experienced "from within" '.[1] To a large degree such a project – and the retrievals of many other neglected geographies of the rural and beyond – revolves around sensitivity to the 'other'. This, of course, in some settings at least, is a sensitivity much sought after and discussed these days (quite rightly), but it is none the less difficult to achieve, and part of that difficulty is seeing how our views of others are constructed and constrained in the very act of viewing and all the baggage of embedded assumptions which accompanies it. The problem is, as Bauman (1993: 91) puts it, 'the Other is recast as my creation; acting on the best of impulses, I have stolen the Other's authority'. This chapter is intended to explore some of the ways in which children, and particularly children in the countryside, are 'seen', by drawing on (mainly) a variety of 'cultural texts'. It ends with a brief consideration of the complexity involved in trying to reach the otherness of children.

VIEWING OUR VIEWING: SEEING STRUCTURE

To see how we see things and the consequences of that seeing is a task beset by difficulties. We can never get 'outside' or 'above' for some form of total view; rather, we are trying to 'map' deep and complex landscapes from a few locations within them. This is surely so in the case of 'country childhoods', for notions of both country and childhood are massively and intricately present within our culture(s). Where they intersect in ideas of 'country childhoods' they often become a vision of considerable potency, bearing a heavy burden of yearning from within the contexts of (adult) modernity. This has the double effect of both being a structuring influence of rural children's lives and, at the same time, disguising the consequence of such and other structuring. Furthermore, such influences spread beyond these immediate settings, and affect how we perceive and structure other childhoods, particularly those set in urban contexts. For as Ward (1990) points out from his perspective of extensive involvement in considering the relationships between children and their environments, urban child-hoods are often prejudged against underlying notions of country childhood idyll, and the contemporary lives of rural children themselves are lived within the shadows of the figures of children that play throughout the sunlit landscapes of popular and literary imaginations.

Ward's instructive exploration of such themes became the springboard for Philo's (1992) pivotal 'Neglected rural geographies', in which the worlds that Ward explores are held up to be ones that have been largely closed to established (modern) rural studies. Philo argued that here was a sensitive, lucid illumination of other worlds which should be of central interest to rural geography, especially given various theoretical developments such as the 'cultural turn' and postmodernism, and their concomitant concern for others, which were providing a powerful pull towards such approaches. Philo also posited that there were many more rural worlds which were suffering similar neglect under the gaze of established modernist practices. In the context of rural childhood, Philo emphasised that rural geography should be particularly concerned with how children's worlds are 'structured "from without" and experienced "from within" ' (p. 198), and tied this into a wider awakening within geography to the generally under-studied status of children, as expressed by James (1990, 1991) and Sibley (1991).

Underpinning much of the structuring of children's worlds in the countryside are various bodies of discourse which portray childhood in the countryside as some form of ideal. Discourse here is taken as Gregory (1994: 11) termed it – 'not just another word for conversation . . . it refers to all the ways in which we communicate with one another, to that vast network of signs, symbols, and practices through which we make our world(s) meaningful to ourselves and others'. These are picked up and translated into the practice of everyday individual, family and institutional practice, and, in the processes of such, feed back into and become part of the ongoing fabric of interconnecting discourses, such as popular, pro-fessional *and* academic (Jones 1995). The structuring which crystallises out of such discourses comes in many interacting, deliberate and incidental forms, and will shape both spatial and temporal patterns of everyday life through physical, ideological, aesthetic, disciplinary and creative forces. Its outworking will also come in many forms and can best be illustrated by some examples. It can range from very fundamental, deliberate structuring of children's lives to the extent of parents/families moving to the country-side specifically to give their children a 'country childhood'[2] to the more everyday 'personal geographies of childhood' in which the 'boundaries' and 'those markers we use to carve up time' (Sibley 1995) are in part drawn in the context of 'country childhood' discourses. Beyond these, other structuring such as play space provision, welfare provision, and all the aspects of how the state intersects with the personal will be shaded by such assumptions and 'readings' of country childhoods. Of course, such discourses of country childhood idyll, although powerful, are not hegemonic, and there are in fact a number of counter discourses (and I will briefly touch on some of these), but these are often set in terms of the (possible) problems that rural children (might) face being overlooked *precisely because* of the pervasiveness of ideas of idyll.

As I have already hinted, ideas of the country (and rural) and of childhood are extensive and complex in both differentiation and contestation. So to attempt to unpick such discourses and reveal their anatomy and implications in their entirety would be futile. Indeed, such is the incoherent and shifting nature of these discourses that to claim them to be a clear entity would be an oversimplification, and, to compound such complexities further, no necessarily objective viewpoint can be claimed; such discourses will have different realities for different perspectives. But having said that, I believe it is possible to pick out (some) major themes within such discourses and to identify icons which stand out within them. So here I aim to do no more than present fragments of discourse and views of icons which I feel show some of the range and depth of ideas of country childhoods within our culture(s), some of their distinguishing elements, and some of their heritages and legacies.

STORIES ABOUT COUNTRY CHILDHOODS

Country childhood discourses come in many shapes and sizes. They can, for example, range from images on packaging and promotional material, through articles in rural lifestyle magazines, to some of the most popular and prevailing works of English literature. Books such as Laurie Lee's *Cider with Rosie* and Flora Thompson's *Lark Rise to Candleford* not only are powerful evocations of recollected rural childhoods but become assimilated into the ongoing development of discourses as they are re-read by subsequent generations. These are just two obvious examples of a whole body of 'rural writing' (Keith 1975) which consists of fiction, autobiography and poetry in which the celebration of country childhoods is a key theme. Ward (1990) feels that such texts have played a key part in the development of country childhood 'stereotypes', for they are 'intensely literary in origin' (p. 18).

To turn to another arena of discourse, Shurmer-Smith and Hannam (1994) point out that there has been a 'marked proliferation of rural lifestyle magazines' (p. 196), which can be seen as indicative of the powerful cultural capital invested in images of the rural in the British context. Within these discourses of rural idyll a subtheme of rural childhood idyll is evident.

For example, the magazine *Country Living* – whose spine always bears the legend 'When Your Heart Is in the Country' – since its launch in the summer of 1985 has carried in each monthly edition the regular feature 'My country childhood' in which the great and the good (and the not so great and not so good) reminisce about their country childhoods.

STORIES FOR CHILDHOOD

But as well as these stories *about* country childhoods there is also a massive wealth of stories (and products) *for* childhood with the countryside and nature as their theme and/or setting. (See Bunce (1994: 63–8) for a review of the role of such material in creating 'the countryside ideal'.) Such material reveals the deep seam of positive association adults often hold between children, nature and the countryside. Such stories reflect not only how children are seen, but also how the countryside is seen. These views and associations become explicitly and forcefully encoded in processes of consumption and commodification, as the stories and products are conceived, produced and marketed by adults and are targeted at children *and the adults* who will deliver them through purchasing, giving, reading and planned playing, etc. Such processes in the context of children's toys and the production and consumption of images of the rural idyll are explored by Houlton and Short (1995).

Hunt (1995) sees the rural as so prominent in children's literature that he concludes that a history of children's literature reveals 'common elements in what the adult sees as the essential children's book . . . a potent mixture of nostalgia (often in the form of a *rural* or suburban *arcadia*)' (p. xi, emphasis added). To think of such classics as A. A. Milne's *Winnie-the-Pooh* stories, the Beatrix Potter stories, Kenneth Grahame's *Wind in the Willows*, Arthur Ransome's *Swallows and Amazons*, Enid Blyton's *Famous Five* series and Richmal Crompton's *William* stories is to think of rural landscapes – landscapes which are often specifically articulated through maps and illustrations within the texts. Although these stories vary in type and in the ages of children they are aimed at, the (mostly idyllic) countryside is ubiquitous among them. Such products cannot be seen merely as discourses from earlier eras, for today they are still aggressively marketed with prominent point-of-sale units in many bookshops, and are also developed and adapted in many new forms such as television and video adaptations, sequels and merchandising. Also, new books and television or video products which are set in rural or countryside settings, such as *Postman Pat* and the *Creatures of Farthing Wood*, have emerged as some of the most successful of more recent children's stories series.

COUNTRY CHILDHOOD IDYLLS

The country childhoods portrayed in stories such as those mentioned above, and the associations between childhood and the countryside (and nature) more generally, reveal just some of the ways in which 'country children', children, the country and nature are 'seen'. My re-reading of these leads me to posit that country childhoods are seen powerfully in terms of a synthesis of innocence, wildness, play, adventure, the companionship of other children, contact with nature, agricultural spaces and practices, healthiness, spatial freedom, and freedom from adult surveillance. To illustrate, many of these are present when Laurie Lee in *Cider with Rosie* recalls the summer night games of his childhood:

> When darkness fell and the huge moon rose, we stirred to a second life. Then boys went calling along the roads, wild slit-eyed animal call, Walt Kerry's naked nasal yodel, Boney's jackal scream. As soon as we heard them we crept outdoors, out of our stifling bedrooms, stepped out into the moonlight warm as the sun to join our chalk-white, moon-masked gang.
>
> Games in the moon. Games of pursuit and capture. Games that the night demanded. Best of all Fox and Hounds – go where you like, and the whole of the valley to hunt through. Two chosen boys loped away through the trees and were immediately swallowed in shadow. We gave them five minutes, then set out after them. They had churchyard, farmyard, barns, quarries, hilltops, and woods to run to. They had all night, and the whole of the moon and five miles of country to hide in . . .
>
> Padding softly, we ran under the melting stars, through sharp garlic woods, through blue blazed fields, following the scent by the game's one rule, the question and answer cry. Every so often, panting for breath, we pause to check on our quarry. Bullet heads lifted, teeth shone in the moon.
>
> 'Whistle-or-'OLLER! Or-we-shall-not FOLLER!' It was a cry on two notes, prolonged. From the other side of the hill, above the white fields of mist, the faint fox-cry came back. We were off again then, through the waking night, among sleepless owls and badgers, while our quarry slipped off into another parish and would not be found for hours.
>
> (Lee 1962: 153)

As a starting point, country childhoods, as in the above extract, are mostly seen in terms of being outdoors. The poet Edward Thomas in his account of his own childhood (1938) tells how the last two sentences of Richard Jefferies' *The Amateur Poacher* – 'Let us get out of these indoor modern days, whose twelve hours somehow have become shortened, into

162

the sunlight and the pure wind. A something that the ancients thought divine can be found and felt there still' – became for him 'a gospel, an incantation' (p. 134). Such sentiments reverberate through ideas of country childhoods, though it should be added that the outdoors seems to be associated with childhood in general. Titman (1994), drawing on the work of Chawla (1986), who explored adult memory of childhood interaction with place, points out that 'some of their strongest memories and recollections of childhood relate to places, often 'outdoor places' (p. 6), and these 'outdoor places were remembered out of all proportion to the relative number of hours spent there' (p. 7). It is inevitable that if the opportunity of children to be outside is restricted or eliminated (and there is increasing evidence of curfews being imposed on children owing to parental fear), or cultural changes to childhood turn children indoors (which again is suspected and feared), then the whole notion of 'a country childhood', and other (outdoor) childhoods, becomes removed still further from the lived experiences of children.

In the extract from Lee, once outdoors the children of the village gather together, and this reflects another prominent idea of country childhoods, that of children being, playing or having adventures *together*, beyond the surveillance and interference of adults – be it the gangs of children's literature already mentioned (*The Famous Five, Swallows and Amazons, William* and his friends 'the Outlaws'), or family groups such as E. E. Nesbit's *The Railway Children*, or, in autobiographical accounts, the children of the village as in *Cider with Rosie*, or, again, sets of siblings as in Herbert Read's (1933) account of his life growing up on a farm. Like being outdoors, such ideas of companionship reflect more general perceptions of childhood, and this is caught in Roberts' (1980) observation that Rousseau's 'fictional child "Emile" signally lacked the most important element of play – other children' (p. 41). Again the implications here are obvious if Ward's (1990) observation, reiterated by Philo (1992: 197), that 'today's rural landscape has fewer children' is true. For if changing demographic and social profiles of rural populations do mean reduced numbers of children in the countryside, for those remaining 'the idyll' might become a site of isolation and loneliness.

Once outdoors (and mostly in company), country children are seen as being blessed through their proximity to and interaction with nature. Accounts such as Flora Thompson's *Lark Rise to Candleford* are shot through with quite detailed and celebratory descriptions of the flora and fauna the children encounter. Nature as spectacle, as provider of interests, hobbies, play spaces, play props and currencies, is seen as bringing not only joy and entertainment but also physical and spiritual health. But the positive association which appears to exist between children and nature – and the countryside as surrogate nature – is complex and deeply rooted within our culture(s), and needs much more exploration than is possible here. However, I will touch upon the critical notions of innocence and naturalness.

The association between children and nature and consequently the countryside can be seen as deeply shaded by the legacies of Romanticism, which, in some of its earliest and most notable expressions, such as Blake's *Songs of Innocence*, synthesised ideas of childhood and nature/countryside into a vision of innocence which was to be juxtaposed against the Enlightenment's deadening traits of experience, adulthood, urbanity and rationality. But it was Rousseau's *Emile* which, according to Brown (1993), had far-reaching influences on the literary portrayal of the child in England, and on educational theories. Here Rousseau saw Enlightenment knowledge and rationality as corrupting 'the natural man', destroying the state of primitive innocence, and to counter this the child Emile is raised in 'rural seclusion' (ibid.) under 'a new scheme of education in which [he is] allowed full scope for individual development in natural surroundings, shielded from the harmful influences of civilisation' (Drabble and Stringer 1985: 851). This idea that the innocence of children is best preserved and expressed within the innocence of nature/the countryside is accompanied by the notion of children being seen as innocent, or in a state of naiveté, in the face of their own mortality. Again this could be said to be a characteristic of adult perceptions of childhood in general, but it is the countryside which is seen as the harmonious environment which is the least likely to disrupt this happy state, or the time 'before I knew I was happy', as Dylan Thomas called it (1965: 52). Retrieving this time is one of the key themes of Thomas's 'Fern Hill' and 'Poem in October', which are considered among his highest achievements (Tremleth 1991). As the first three stanzas of 'Fern Hill' show, such nostalgic adult yearning can become an overwhelming part-evocation, part-lamentation for a country childhood remembered as lived out among nature and agricultural spaces:

> Now as I was young and easy under the apple boughs
> About the lilting house and happy as the grass was green,
> The night above the dingle starry,
> Time let me hail and climb
> Golden in the heydays of his eyes,
> And honoured among wagons I was prince of the apple towns
> And once below a time I lordly had the trees and leaves
> Trail with daisies and barley
> own the rivers of windfall light.
>
> And I was green and carefree, famous among the barns
> About the happy yard and singing as the farm was home
> In the sum that is young once only,
> Time let me play and be
> Golden in the mercy of his means,

And green and golden I was huntsman and herdsman, the calves
Sang to my horn, the foxes on the hills barked clear and cold,
And the sabbath rang slowly
In the pebbles of the holy streams.

All the sun long it was running, it was lovely, the hay
Fields high as the house, the tunes from the chimneys, it was air
And playing lovely and watery
And fire green as grass.
And nightly under the simple stars
As I rode to sleep the owls were bearing the farm away,
All the moon long I heard, blessed among stables, the nightjars
Flying with the ricks, and the horses
Flashing into the dark.

(Thomas 1952: 159–60)

Such literary and philosophical views of the innocence of children being at home in the countryside can find echoes in more everyday interpretations. Lauren Young (*She* magazine, August 1995), when recounting her family's move from London to rural Dorset, stated, 'there's an innocence and sweetness about country children that our eldest son had almost lost' and that two years after the move he was 'quite transformed'. In the context of my research one mother told me that living in a country village 'had prolonged' her children's childhood in that they had retained a state of innocence longer than she felt they would have done if they had lived in an urban location.

In some ways, the countryside itself has been seen as innocent and therefore other to modernity, for, as Turner (1994) argues, the origins of modernity are 'associated necessarily with the development of an industrial *urban* capitalist society' (p. 21; emphasis added). 'Rural writers' such as Massingham (1939) and Batsford (1940) wrote passionately about how they saw modernity spreading from the cities and overwhelming the traditional beauty, variety and innocence of the countryside. Thus the coming together of childhood and the countryside both confirms and enhances these states of innocence within the face of rational (urban) modernity. Now in these days of massive concern over the status and condition of childhood, perhaps the countryside is becoming the last refuge of ideas of childhood idyll. Images of inner-city crisis often contain children, and images of childhood crisis are nearly always set in urban contexts. In one of the most traumatic incidents, which has quickly become a standard reference of childhood crisis, the toddler James Bulger was seen in grainy store security videos being led from a shopping mall to his death at the hands of two other children. The wildness of (younger) children in the countryside is often seen as innocent and wholesome and mostly celebrated, but wild children in the city are seen as feral, even vermin, as the story of 'Ratboy' indicates.[3] (And indeed, in other urban settings they are treated and exterminated as such.)

Country childhoods are seen not only in terms of contact with nature but also as nature mediated through agricultural processes and other charms of agricultural landscapes. Part of this complex synthesis and its perceived benefits is captured in the preface to Munro's (1978) assessment of 'the rural child', who is

> an individual, secure and important in his own right ... is closer to the natural world – of growing things, seedtime and harvest, decay and renewal. Because of this he is quieter and usually more serene – somehow slower yet often more assured. He is aware of belonging to something real, and what he sees in the countryside reassures him to this knowledge.

Many of the evocations of country childhoods are set in agricultural communities and landscapes, and they draw upon the differing nature of agricultural spaces, forms and surfaces. The fields, hedges, woods, farmyards, farm buildings, animals, processes and technologies of farming are all deeply present in accounts of country childhoods (see, for example, Davies 1989; Read 1933). In *Cider with Rosie*, Lee vividly recounts how on one winter's day he and other children from his village spontaneously decide to visit one of the local farms, where

> Inside the cowsheds it was warm and voluptuous, smelling sweetly of milky breath, of heaving hides, green dung, and udders, of steam and fermentations. We carried cut hay from the heart of the rick, packed tight as tobacco flake, with grass and wild flowers juicily fossilised within – a whole summer embalmed in our arms.
>
> I took a bucket of milk to feed a calf. I opened its mouth like a hot wet orchid. It began to suck at my fingers, gurgling in its throat and raising its long-lashed eyes
>
> (Lee 1962: 139)

There can be little doubt that British agriculture has gone through a number of radical changes in its practices, internal structures and its relationship with the countryside and rurality. Nearly all the accounts of country childhoods used here are set within agricultural landscapes, processes and structures which have long since gone through such radical change. Even the most recent accounts, because they are being written by adults remembering childhoods, are recalling scenes of at least twenty years ago. In fact, most of the iconic accounts are from earlier in this century or even from the late nineteenth century, so there is an inevitable lag between the images of childhood which emerge and are sustained by such accounts and the reality of children's lives in contemporary agricultural landscapes. This, and other changes the countryside and childhood have gone through, leads to the need to unravel the questions – was the notion of childhood countryside idyll some form of 'reality' which has now been eroded by changes to the countryside and/or to childhood?; or has it been transformed into new forms?; or was/is it largely a creation of adult nostalgia which never had much grounding in the everyday lives of rural children?

When Shoard (1980: 192) states that 'the countryside is of course a boundless treasure-house of opportunities for creative play, and one for which no real substitute has ever been found', she is confident that the idyll was rooted in some form of reality and that it is now under siege. She sees the countryside as the ideal territory for childhood, and implies that other territories, presumably including urban ones, where the vast majority of children now live, are second best. She goes on to support her stance by recounting a day spent with children from the village of Minster, who were 'adamant that the country was a much better place to play in than the street, the playground, the recreation ground with its grass, swings and slide, or even the seaside' (ibid.). Shoard then analyses what it is that the countryside offers children in terms of play, breaking it down into five categories:

> First, it provides a much greater variety of 'props', from bullrushes to spiders, long grasses (musical instruments as well as craft material) to snails. Second, the countryside provides much greater freedom of movement: not only does long grass make possible games like 'Soldiers': it also enables children to feel safe. They believe they can take a tumble while playing leapfrog, for instance, without hurting themselves. Third, the countryside provides a separate place to which they can retreat from their homes, or rather their parents' homes, in the village. The construction of little dens or miniature houses is a much favoured activity. Fourth, the animals of the countryside provide a rich source of enjoyment – watching birds, catching tadpoles, collecting spiders to frighten parents with, racing snails, letting grass snakes slither through the fingers, letting frogs bounce up and down in the hand. Finally, the countryside is a source of the unknown: it provides endless surprises and makes possible unexpected scrapes. Because there is so much to discover in the country, many of the children I spoke to were happy to spend time simply looking around, finding new places, stumbling across exciting, unusual things, getting into scrapes, skipping from one pursuit to another. Whereas woods, for example, are ideal for anything from playing hide-and-seek to chasing animals or playing stunts on bicycles, 'at the seaside all there is to do is swim and walk along the beach and make sandcastles'.
>
> (Shoard 1980: 193)

Shoard goes on to recount how the children, in the past, had played mainly in the areas of marshes near the village to which they had *de facto* access. Subsequently, much of this had been reclaimed for intensive agriculture, and the children now had to rely on 'several scraps of rough marginal land' which still remained. She illustrates her concern that the opportunities rural children have for play, and all the benefits it brings with it, are being eroded, with interview extracts from children whose favourite play places had been grubbed out or ploughed up in the processes of agricultural intensification, and concludes, 'across the countryside, this kind of devastation must already have changed the experience of childhood in England substantially' (ibid.). So for Shoard the country childhood idyll is, or at least was, a reality, and it was enabled by the physical attributes of the landscape, but changes to the landscape are cumulatively reducing such opportunity.

Shoard is passionate and eloquent in her polemic against the 'theft of the countryside', but apart from her obviously genuine concern about the consequences of countryside change and their impact on various social groups, including children, the latter are also a usefully emotive factor to deploy in her overall argument. As is so often the case, this sort of story of lost, or about to be lost, golden ages has been told many times before, and this was an issue explored by Williams (1985) within his now famous analysis of images of the city and the country. At one point he considers the poet Clare's lamentations for the loss of the countryside of his childhood in the enclosures and other agricultural developments of the eighteenth century, and sets out to show how these stories are more complex than tales of purely material landscape change.

That material change did and still does occur is not in doubt, and Williams extracts from the work of Clare the fact that the 'primitive land' was 'being directly altered: the brooks diverted, the willows felled, in drainage and clearance' (Williams 1985: 138) (much the same sort of land change that Shoard refers to). But Williams adds that 'particular trees, and a particular brook, by which I played as a child, has gone in just that way, in the last few years, in an improved use of marginal land' (ibid.), and he goes on to explore how the landscapes that Clare laments were the landscapes of his childhood, and it becomes uncertain which is being mourned. The joys of childhood are imprinted onto whatever landscape they are acted out on. The process of adult recollection tends to convert such particular memories into 'sweet vision of the past'. Thus 'the natural images of this Eden of childhood seem to compel a particular connection, at the very moment of their widest generality. Nature, the past and childhood are temporarily but powerfully fused' (ibid.: 139).

Although such analysis casts doubt on the straightforwardness of adult evocations of childhood idylls and lamentations for idylls lost, support can be found for Shoard's view. Humphries *et al.* (1988: 63) consider that 'there is some truth' in the evocations of 'natural' play in accounts such as *Cider with Rosie* and *Lark Rise to Candleford*:

> For despite the fact that most country children had no money and few toys, they could enjoy climbing, fishing, hunting and picking wild fruit – all for free. The migration to the cities, and decades of agricultural depression, had left a picturesque landscape with many a ruined barn and rambling hedge for children to play in. In rural areas the countryside became, in children's eyes, a great adventure playground where their imagination could run wild, building tree dens, damming streams and raiding birds' nests.

But despite concerns that today's countryside is less amenable to children and their activities, – 'there are few places for children to play in the countryside'; National Children's Play and Recreation Unit (1992: 4) – country childhood idylls still hold powerful sway. Evidence of the charms farms and farmyards are still seen to have for children is evident in instances of their commercialisation as farmers use them as a means of diversifying out of the agricultural crisis in the UK. A travel article on such ventures (*Observer*, 14 March 1993) featured Treffery Farm, Bodmin, Cornwall, reporting, 'children will enjoy this 200-acre dairy farm where hens and ducks wander freely. They can watch the milking, help feed the calves, collect the eggs and ride Fred the docile pony'.

When does play become adventure? Perhaps all play is to an extent an adventure, but, as indicated earlier in the 'Stories for childhood' section, the countryside becomes the place of adventure in childhood literature. The very titles of Enid Blyton's *Famous Five* adventure stories reveal their dependence on access to countryside: *Five: – On Finniston Farm, Go to Smuggler's Top, Go Off in a Caravan, On Kirren Island Again, Go Off to Camp, Go to Mystery Moor, Go to Billycock Hill Go to Demon's Rocks*.

Other equally famous adventure stories, such as Ransome's *Swallows and Amazons*, are again set in 'real' landscapes and 'real' events but onto these physical adventures of rival groups of children sailing and camping are overlaid imaginative reinterpretations where the children (or the author on their behalf) refictionalise their adventures into other narratives of pirates and treasure. This is a particularly strong feature of Richard Jefferies' *Bevis*, where the adventures within a small area of wood and scrubland, and on the stream running through it, become within the children's minds truly global as they explore 'The Mississippi', 'By the New Nile', 'Central Africa', 'The Jungle', and encounter 'Savages'. Yet others still combine physical and imaginative adventure with supernatural, magical and historical elements, notable examples of these being Rudyard Kipling's *Puck of Pook's Hill* and John Masefield's *The Box of Delights*, and more recently the work of Alan Garner has continued these traditions with works such as *The Owl Service*. Critically, it is the countryside which is chosen as the settings for all these works, for it provides not only the fictional physical space for adventure, but also the imaginative, even mystical space in which such narratives can unfold unimpinged by the jarring nature of much of everyday life. This is a powerful source of 'viewing the countryside' for the children and adults who read such texts and view the many television adaptations of such and other works.

As well as the countryside being seen as an ideal place for play and adventure for children, there is also an ongoing and powerful suite of ideas revolving around its being a place of health and healing for children, both in physical and psychological terms. Perhaps the most notable fictional rendering of such ideas is F. H. Burnett's *The Secret Garden*, one of the great icons of children's literature (Gunther 1994: 159), in which the central character arrived as a 'yellow faced, sickly, bored and wretched child' but

> when her mind gradually filled itself with robins, and moorland cottages crowded with children ... with springtime and with secret gardens coming alive day by day, and also a moor boy and his 'creatures' there was no room left for the disagreeable thoughts which affected her liver and her digestion and made her yellow and tired.
>
> (Burnett 1993: 283–4)

Both the heroine and her companion, Colin, who was a 'hysterical, half-crazed little hypochondriac who knew nothing about the sunshine and spring' (ibid.), were regenerated, made wholesome and healthy by the contact with the nature, the countryside and its inhabitants.

Such attitudes can be seen re-emerging in a number of different forms. In the mass evacuation of children from London and other cities in the Second World War to rural areas considered safe from bombing, there was at once immediate concern over the state of health many of the city children were found to be in, and also an assumption that the spell in the countryside would do them good. Macnicol (1986) points out that the authorities organising the evacuation had a vested interest in such a prognosis, namely to discourage 'drift-back to the danger areas', but he also adds, 'it seemed tantalisingly obvious that country life was healthier' and that 'everybody expected the urban child to benefit' (p. 19). But as it turned out, the 'urban children suddenly subjected to the apparently healthy environment of fresh air and green fields did *not* show any marked improvement in health' (p. 18). Today such ideas are still embedded in attempts to reunite urban children with some aspects of the countryside. The mission of the charity Farms for City Children is to take groups of children 'from the classrooms of the bleakest inner-city streets' and 'set them down in the classic agricultural landscapes of South-West England' (*Country Living*, October 1993). Conversely, the City Farm movement provides places which are 'sited in some of the most neglected parts of our towns and cities' where 'inner city children are able to learn first hand how to look after animals and plants, how farming produces food for the table' (current publicity leaflet). The City Farm movement in fact has a multifaceted and admirable agenda, but at the core of its ethos and image – as evident in the pictures of children with animals which illustrate its publicity – is the idea that city children are missing something key to childhood.

COUNTRY CHILDHOOD SPACES

Both James (1990, 1991) and Sibley (1991) are concerned with how children live in a world which is mostly physically ordered and scaled on adult terms. They advocate that geography should be sensitive to both the structuring this imposes and the resistance and subversion it meets. Such spatialities of childhood come across strongly in accounts of country childhood idylls, and it is instructive to see the forms these take. For within the spatiality of these adult recollections of country childhoods may lie scope for unravelling the extent to which these accounts were and are in touch with children's lived experiences of the rural, or otherwise reflect processes of adult idealisation and imagination.

In *Bevis* by Jefferies (1882), almost all the action of the vast three-volume story is set in a small terrain which is described in the opening pages:

> The brook made a sharp turn round the withy-bed, enclosing a tongue of ground which was called in the house at home the peninsula, because of its shape and being surrounded on three sides by water. This piece of land, which was not all withy, but partly open and partly copse, was Bevis's own territory, his own peculiar property, over which he was autocrat and king.
>
> (Jefferies 1882: 2)

This extract is one of many examples in which separate play space is seen to play a key role in country childhood idylls. These spaces are separate in so far as they are apart from adult worlds, adult-dominated space. In *Bevis* this space is provided by its being redundant to adult uses; it was thus mostly an adult no-go area. Similarly, in the *William* stories the 'headquarters' of the 'Outlaws', and the pivot to many narratives, is 'the old barn', space from which adult usage again had withdrawn. Critically such spaces needed to be free not only of adult presence but also of adult surveillance, like a favourite place Lee describes in *Cider with Rosie*: 'a good place to be at any time. . . . It seemed down here no disasters could happen, that nothing could touch us. This was Sammy's and Sixpence's; the place past the sheep wash, the hide-out unspoiled by authority' (Lee 1962: 151, 152).

Such spaces were not just found spaces but also created spaces in the forms of dens and tree houses etc. Ward (1990) argues that such building

171

activities are 'the most universal of all children's pretend games' (p. 88), and they provide vital places of privacy for children. Ward goes on to redescribe a number of accounts of children building dens which appear in various texts, with the extremely detailed account in *Bevis* being 'the most famous of all evocations of childhood manipulation of space' (p. 89). Even this aspect of country childhood has been processed into glossy rural lifestyle terms. Beer, in *Country Living* (September 1994), celebrates the tree house, advocating that tree houses should be 'well hidden and parent-proof' yet also providing 'how to build a tree house' tips. It is the country-side (especially when contrasted with the urban) which is seen as providing the greatest opportunity in terms of spaces, materials and access to both, for children to undertake such activities.

Other separate spaces are created by temporal structuring, when adult use of a space is intermittent, and in the time when the space is not in use, children can move in and take over. There are many examples of these kinds of spaces within accounts of country childhoods, and often these were generated by the seasonal, and even daily cycles of agriculture. Barns, cowsheds, fields often became temporarily redundant spaces where the activities of children would not come into conflict with other uses. Harvest time is a classic example of this, and it does feature in many accounts of country childhood (e.g. Davies 1989), where fields of standing crops were sternly protected from the trampling feet of playing children, but, once cut, what remained were fields of stubble onto which children could venture.

If Shoard's concern, recounted earlier – and that of Ward, that the countryside is being tidied up – is justified, what is under threat is such separate spaces. In part, the countryside was seen as such an idyll because there were such spaces for children, but now these may be dwindling under processes of physical, temporal and ideological homogenisation in which (adult) order becomes ubiquitous. But such separate spaces should not be seen in strictly spatial terms alone. Spaces for children are also seen as separate in their interpretation by children. Frequently adults and children are seen as using and seeing the same spaces but putting differing interpretations on them, thus creating the parallel worlds of adulthood and childhood (Roberts 1980). These spatial and imaginative separations were often synthesised together, as is captured by Kenneth Grahame in the prologue to *The Golden Age* in which he contemplates adults – called Olympians – from a child's point of view.

> On the whole, the existence of these Olympians seemed to be entirely void of interests, even as their movements were confined and slow, and their habits stereotyped and senseless. To anything but appearances they were blind. For them the orchard (a place elf-haunted, wonderful!) simply produced so many apples and cherries; or it didn't

– when the failures of Nature were not infrequently ascribed to us. They never set foot within fir-wood or hazel-copse, nor dreamt of the marvels hid therein. The mysterious sources, sources as of old Nile, that fed the duck-pond had no magic for them. They were unaware of Indians, nor recked they anything of bisons or of pirates (with pistols) though the whole place swarmed with such portents. They cared not to explore for robber's caves, to dig for hidden treasure. Perhaps indeed, it was one of their best qualities that they spent the greater part of their time stuffily indoors.

(Grahame 1928: 2)

This (now somewhat dated) extract demonstrates how complex the layering of adult constructions of childhood can be. What it actually consists of is an adult construction of how children see adults seeing children. It also shows that some adult versions of the country childhood idyll have seemingly already built into them children's subversion of space, children's 'otherness', by adapting the child's point of view within the narrative, and this adds yet another dimension to adult constructions of country childhood idylls which has to be unravelled.

Throughout many accounts of country childhoods there are strains of 'anti-order'. These should not necessarily be seen as deliberate calls to rebellion but more as expressions of what is seen as the innate contradiction between adult order and the 'disorder' of children. In spatial terms, many of the accounts of country childhoods highlight not only derelict or abandoned spaces but also untidy corners and spaces where the adult control of the space is more conducive for children precisely because the 'normal' adult ordering has not taken place. Dylan Thomas recalled his delight at his uncle's farmyard and how it was the dilapidation itself which made it so attractive to him as a child:

The pigsties were at the far end of the yard. We walked towards them ... past three hens scrabbling the muddy cobbles and a collie with one eye, sleeping with it open. The ramshackle outhouses had tumbling, rotten roofs, jagged holes in their sides, broken shutters, and peeling whitewash; rusty screws ripped out from the dangling, crooked boards; the lean cat of the night before sat snugly between the splintered jaws of bottles, cleaning its face, on top of the rubbish pile that rose triangular and smelling sweet and strong to the level of the riddled cart-house roof. There was nowhere like that farmyard in all the slapdash county, nowhere so poor and grand and dirty as that square of mud and rubbish and bad wood and falling stone, where a bucketful of old and bedraggled hens scratched and laid small eggs. A duck quacked out of the trough in one deserted sty. Now a young man and a curly boy stood staring and sniffing over a wall at a sow, with its tits on the mud, giving suck.

(Thomas 1965: 12)

Possibly linking to this theme of disorder in images of country childhood idylls are the encounters with Gypsies, or ideas of their nomadic lifestyle, which crop up on numerous occasions in accounts of country childhood idylls, and it is interesting to see the associations between childhood 'otherness' and this other form of rural 'otherness' which was once apparently part of the 'rural idyll', but which now seems so under pressure in today's cultural and ideological climates. Williamson (1932), Thompson (1973) and Davies (1989) all give vivid accounts of childhood encounters with gypsies, and in Grahame's *Wind in the Willows*, if not Gypsies themselves, then at least their nomadic existence on the open road is portrayed with relish by Toad as he cajoles Ratty and Mole to join him in his Gypsy caravan holiday. Lurie (1991), when considering the strengths of the children's fiction of Mayne, shows how 'otherness' in some form or other may be a key ingredient to children's literature and, through this, childhood itself. She suggests that Mayne's 'important characters are usually children or innocents, unsophisticated, half-literate people, *separated from the contemporary world in some way* – they are gypsies, servants and labourers, farmers in remote Yorkshire villages or inhabitants of an earlier period' (Lurie 1991: 202; emphasis added). So perhaps it is in these themes of spatial and cultural disordering that there is a sensitivity to the otherness of children, for, as the Opies reported (Opie and Opie 1969) in the context of their famous surveys of children's games and folklores, 'the peaks of a child's experience are . . . occasions when he escapes into places that are disused, overgrown and silent' (p. 15). But of course all these are still adult constructions of childhood, and therefore need to be treated with caution in respect of assuming that they have truly entered children's worlds.

DIFFERENTIATION, IDEALISATION, CONTESTATION AND REPRESENTATION: A SORT OF CONCLUSION

In this chapter I have tried to provide some glimpses into the many ways in which children in the countryside are seen, and the assumptions which inform such ways of seeing. In the short space provided, this has become little more than a generalising overview of ideas of country childhood idylls. There has been little room for dealing with the differentiation within such notions which would be expected to follow various differences such as children's age, gender, social/cultural contexts, etc.,[4] and also, differing types of 'rural'. These different versions and *applications* of childhood idylls warrant detailed study, although I do suggest that they may be similarly tinted by overarching, generalised discourses of positive association between children and the countryside, which would have to be confronted within such work.

As already posited, how rural (and other) children's worlds are structured to a considerable extent stems from adult notions of country childhood idylls, and one of the tasks facing those trying to unravel such structuring is trying to unpick to what extent the countryside is some sort of idyll for children which is merely *represented* by adults or whether it is an idealisation *constructed* by adults, or – given the probable messy interpenetration of such poles – which bits are which and how they interact. Within this, the passage of time seems to play a critical and complex role. Not only is the task to understand how rural spaces and childhoods have changed and the impacts of those changes, but also how such spaces and childhoods are remembered and seen. If, as often suggested (e.g. Short 1992), the countryside (and children in it) is seen through a process of nostalgic remembrance, 'the material countryside' of today is seen blended with past 'material countrysides' which themselves are made imaginary through remembrance and idealisation. We are faced with a labyrinthine hermeneutic of interpretations of interpretations of temporally and spatially jumbled landscapes and the lives within them.

I have also tried to touch upon various contestations which aim to challenge the idealisation of country childhoods, and which instead seek to represent as transparently as possible the experiences of children who have lived or do live in the countryside. Some of these surely do contain degrees of sensitivity to the otherness of children, especially those which are initially orientated around issues of children's welfare and rights.

But even in accounts of country childhood idylls, there are fleeting glimpses of other sorts of country childhoods. Literally this is so in *Cider with Rosie*, when the children, calling on one of the local farms, briefly see a sickly child who waves to them from his window, and who 'wouldn't live to last out the winter'. In the influential *Bevis* there is a scene which is a sharp reminder that the opportunity for children to share in the countryside idyll was often dependent on class status. Bevis, in his seemingly endless free time, has made a raft to sail on the stream, but, having problems launching it, calls to one of the farm employees, a boy working nearby, to help. The boy is reluctant but Bevis petulantly insists, and together they get the task done. The boy begins to enjoy himself and forgets his duties, until he 'felt a grip on the back of his neck', and the farm bailiff 'marched him up the meadow'. Meanwhile 'Bevis (and Mark) were too full of the raft to even notice that their assistant had been hauled off' (Jefferies 1882: 20). While Bevis is free to play, other children were working on his father's farm, and Humphries, *et al.* (1988) stress that children of farm labourers and poorer farmers had very little time for play, being put to work as soon as they were physically able to. In these two examples differences in health and wealth or status mean that some country children can be seen as living lives other than idyllic even within the very texts which are, or have been, icons within discourses of childhood idyll. Many other forms of contestations can be identified by such means, and these can be taken as indicators of other stories being lived out by children in rural places today which may need telling as contestations of the idea of idyll. But it should be noted that the well-intended desire to do so must not become merely a counter-idyll discourse, for such versions of child-hood may still be too firmly tied into yet another (adult) way of seeing children.

Trying to bring all these ways of seeing children and, particularly, dominant cultural discourses into some sort of reflexive critical context is, I feel, vital for enquiring into how children's worlds are structured, and for going as far as possible 'towards' the otherness of children. We have to try to understand how we may be imposing 'otherness' rather than be open to it. And if the assertion that 'we have pinned onto children as individuals, children as persons, a whole enormous philosophical edifice, about something called childhood, which is not at all what the condition of children is' (BBC 2 *Late Show*, 5 December 1994) has any credence, then efforts to reach the 'other/other' of 'children', rather than the 'other/same', are truly challenging (there terms are explained by Chris Philo in Chapter 2 of this volume).

We have all been 'children', or at least biologically young, so (perhaps uniquely in this concern for a form of otherness) we have all been that other once, and may still contain some form or traces of it. This raises the question of whether it, or elements of it, are retrievable through memory, or whether the *illusion* that it is retrievable in fact makes the other/other even more inaccessible and invisible. Children have different views of the world which are filtered through their own stocks of knowledge (Thrift 1985). Once superseded by adult stocks of knowledge, those adult filters can never be removed to get back to earlier states. Adult constructions and memories of what it is or was to be a child are inevitably processed through adultness. From adult perspectives, children's geographies may well appear bizarre and irrational, and the challenge is to translate these into the rational language of academic research and writing, without in the process losing these very characteristics which may be at the centre of understanding children's geographies.

Finally, this brings me back to the concern expressed at the outset, that of the over-neatness of the 'structured "from without" and experienced "from within"' conceptualisation. Clearly there is much usefulness in such an approach; it is easy to think of many examples where children's experience of things from within is different from – contesting or subverting – the structures in which they live their lives. But there must be a question about the experiencing from within itself having degrees of structuring, thus rendering a simple binary module much more complex. Such issues need exploring in the contexts of how the mapping and boundaries children encounter (Pile and Thrift 1995: 56) structure the subject which then does the experiencing. These are urgent requirements. Gregory (1994: 124) observes that feminist geography revealed the orthodoxy which was being challenged to be 'a half-human geography'. Now, growing awareness of the need for children's geographies, and the geographies of other others and their intersections, reveals that it was never even that, but rather some much smaller fraction of human geography.

ACKNOWLEDGEMENTS

This chapter is formed from current research funded as an ESRC Post-graduate Training Award. Thanks are due to the editors for comments on earlier drafts.

NOTES

1 As I suggest later, this is perhaps an over-neat division but it is taken here to be useful in setting up initial conceptual and methodological orientations.
2 This is one preliminary finding of a qualitative research project currently being undertaken which is attempting an empirical interrogation of the discourses of country childhood idylls considered in this chapter. The research, which is in the form of a case study of a small village in Avon, south-west England, intends to explore how children's lives are 'structured from without' by studying both popular discourses present in various forms of media, literature, art, etc. and place-specific lay discourses. In particular, attention is being paid to how the adult lay discourses present within the village interact with wider popular images of childhood, and how in turn these become structuring forces which shape and order the village children's lives. Set against this, the research also aims to make contact with how, within these contexts, the village children's lives are 'experienced from within' by trying to uncover some of their own experiences and perceptions of their everyday lives. See Chapter 8, this volume.
3 'Ratboy', as reported in the *Independent* (9 October 1993), was the title given by a national tabloid newspaper to a 14-year-old boy who was arrested by police for burglary on a housing estate after he had been found hiding out in a ventilation shaft. The story, became emblematic of views of feral children and inner-city malaise.
4 One example of gendered structuring in ideas of country childhood which appears to be emerging from my research is the notion that pre-teen 'country girls' are often seen as 'tomboys', – a notion reflected in some children's literature, e.g. 'George' in the *Famous Five* stories, a girl who 'always tried to be a boy'. Tentatively, it would seem that boyhood in the countryside is seen as the natural state of childhood, and country girls are quasi-boys until the sophistication of femininity somehow overwhelms this 'natural' state.

REFERENCES

Batsford, H. (1940) *How to See the Country*, London: Batsford.
Bauman, Z. (1993) *Postmodern Ethics*, Oxford: Blackwell.
Brown, P. (1993) *The Captured World: The Child and Childhood in Nineteenth-Century Woman's Writing in England*, Hemel Hempstead: Harvester Wheatsheaf.

Bunce, M. (1994) *The Countryside Ideal: Anglo-American Images of Landscape*, London: Routledge.

Burnett, F. H. (1993) *The Secret Garden*, Ware: Wordsworth. (First published 1909.)

Chawla, L. (1986) 'The ecology of environmental memory', *Children's Environments Quarterly*, 3 (4): 34–42.

Davies, P. (1989) *Mare's Milk and Wild Honey: A Shropshire Boyhood*, London: André Deutsch.

Drabble, M. and Stringer, J. (1987) *The Concise Oxford Companion to English Literature*, Oxford: Oxford University Press.

Grahame, K. (1928) *The Golden Age*, London: The Bodley Head.

Gregory, D. (1994) *Geographical Imaginations*, Oxford: Blackwell.

Gunther, A. (1994) 'The Secret Garden revisited', *Children's Literature in Education* 25, (3): 159–68.

Houlton, D. and Short, B. (1995) 'Sylvanian families: the production and consumption of a rural community', *Journal of Rural Studies* 11, (4): 367–86.

Humphries, S., Mack, J. and Perks, R. (1988) *A Century of Childhood*, London: Sidgwick and Jackson.

Hunt, P. (1995) *Children's Literature: An Illustrated History*, Oxford: Oxford University Press.

James, S. (1990) 'Is there a "Place" for children in geography?', *Area* 22: 378–83.

—— (1991) 'Children and geography: a reply to David Sibley', *Area* 23: 378–83.

Jefferies, R. (1882) *Bevis: The Story of a Boy*, vol. I, London: Sampson Low, Marston, Searle and Rivington.

Jones, O. (1995) 'Lay discourse of the rural: developments and implications for rural studies', *Journal of Rural Studies* 11, (1): 35–49.

Keith, W. J. (1975) *The Rural Tradition: William Cobbett, Gilbert White, and Other Non-fiction Prose Writers of the English Countryside*, Brighton: Harvester.

Lee, L. (1962) *Cider with Rosie*, Harmondsworth: Penguin.

Lurie, A. (1991) *Not in Front of the Grown-ups: Subversive Children's Literature*, London: Sphere.

Macnicol, J. (1986) 'The effect of the evacuation of schoolchildren on official attitudes to State intervention', in H. L. Smith (ed.) *War and Social Change: British Society in the Second World War*, Manchester: Manchester University Press.

Massingham, H. J. (1939) *The English Countryside*, London: Batsford.

Munro, I. S. (1978) In Preface, *The Country Child*, Lincoln: Centre for the Study of Rural Society.

National Children's Play and Recreation Unit (1992) *Children Today in Devon: Playing in the Countryside – A Study of Rural Children's Services*, London: National Children's Play and Recreation Unit.

Opie, I. and Opie, E. (1969) *Children's Games in Streets and Playgrounds*, London: Oxford University Press.

Philo, C. (1992) 'Neglected rural geographies: a review', *Journal of Rural Studies* 8, (2): 193–207.

Pile, S. and Thrift, N. J. (1995) *Mapping the Subject: Geographies of Cultural Transformations*, London: Routledge.

Read, H. (1933) *The Innocent Eye*, London: Faber and Faber.

Roberts, A. (1980) *Out to Play: The Middle Years of Childhood*, Aberdeen: Aberdeen University Press.

Shoard, M. (1980) *The Theft of the Countryside*, London: Maurice Temple Smith.

Short, B. (ed.) (1992) *The English Rural Community: Image and Analysis*, Cambridge: Cambridge University Press.

Shurmer-Smith, P. and Hannam, K. (1994) *Worlds of Desire, Realms of Power: A Cultural Geography*, London: Edward Arnold.

Sibley, D. (1991) 'Children's geographies: some problems of representation', *Area* 23: 269–70.

—— (1995) 'Families and domestic routines: constructing the boundaries of childhood', in S. Pile and N. Thrift (eds) *Mapping the Subject: Geographies of Cultural Transformation*, London: Routledge.

Thomas, D. (1952) *Collected Poems 1934–1952*, London: J. M. Dent & Sons.

—— (1965) *Portrait of the Artist as a Young Dog*, London: J. M. Dent & Sons. (First published 1940.)

Thomas, E. (1938) *The Childhood of Edward Thomas: A Fragment of Autobiography*, London: Faber and Faber.

Thompson, F. (1973) *Lark Rise to Candleford*, Harmondsworth: Penguin.

Thrift, N. J. (1985) 'Flies and germs: a geography of knowledge', in D. Gregory and J. Urry (eds) *Social Relations and Spatial Structures*, London: Macmillan.

Titman, W. (1994) *Special Places; Special People: The Hidden Curriculum of School Grounds*, Godalming: WWF UK.

Tremleth, G. (1991) *Dylan Thomas*, London: Constable.

Turner, B. S. (1994) Introduction to *Baroque Reason: The Aesthetics of Modernity* by C. Buci-Glucksmann, London: Sage.

Ward, C. (1990) *The Child in the Country*, London: Bedford Square Press.

Williams, R. (1985) *The Country and the City*, London: The Hogarth Press.

Williamson, H. (1932) *The Beautiful Years*, London: Faber and Faber.

179

10

CONTESTING LATER LIFE

Sarah Harper

INTRODUCTION

With the acceptance that space is constructed of a wide range of relations, of economy, of power, of society and of symbol, has come the acknowledgement that it is not only class, race and gender relations that help to locate and determine places, but also other social relations such as sexual preference, age and intergenerational relations. It is the latter – age and intergenerational relations – which will form the focus of this chapter. I shall argue that our knowledge of ageing is gained from a variety of constructed models of late life, drawn from personal and public notions of dependency, autonomy, power, control, body and mind. While such knowledge is disseminated throughout professional and public arenas, and translated into policies to facilitate the ageing experience, rarely is the source of such knowledge questioned. Yet it is from such knowledge that the categories and dependencies of later life are constructed.

As Hazan (1993) has pointed out, behind the rhetoric of such categories lies the assumption that the aged and non-aged constitute two distinct categories of humankind, or to use Gadow's (1986) vivid term, as social objects of special contemporary interest, elderly people have become so special as to be considered a *separate species*. Indeed, I was asked to write a chapter not on ageing but on elderly people; yet elderly people are but adults who in recent years have come to be defined as elderly as a result of certain political acts that have clearly identified the state of old age as commencing at 65 – the formal retirement age for many societies (Harper 1997). It is now fully acknowledged that there are many stages of old age and not all should be associated with notions of deterioration and decay. This however also begs the question: not, why do we confuse the different 'stages' of 'old age', but rather, why do we place all post-65s together in the first place? As has long been realized, the experience of the healthy 65-year-old married male recent retiree is so distinct from that of the frail, widowed woman in her nineties approaching the end of her life (Walker

1981) that to construct both as within the same period of life has little or no analytical credibility.

The chapter thus attempts to examine the contradictions in the confusion surrounding the various constructs of later life. This clearly has relevance for those interested in constructing and contesting the countryside in that many rural areas have a larger than average percentage of their population in later life, conventionally defined as over 65 (OPCS 1995; see Warnes and Ford 1993 for an analysis of this point). As the 1991 UK census revealed, the three rural areas of Wales, East Anglia and the South-West had the highest concentrations of those over 65, with 20 per cent of the population over this age. While this is only slightly higher than the national average of 18.7 per cent, as Warnes and Ford (1993) point out, within the regions certain districts and counties are still more heavily aged, those over-65 reaching 35 per cent of the population in Dorset. Thus, despite the view that retirement in-migration and youth out-migration may well be slowing (Jones 1990), clearly many rural areas are still relatively demographically aged. It is also worth noting that, within the UK, rural retirement migration is restricted among the ethnic minority groups and lower socio-economic groups, as a result of lack of both economic resources and rural-based kin networks (Mutchler and Burr 1991), and though apparently spreading at some level among the working classes, it remains predominantly a middle-class phenomenon (Warnes 1991). While there has been considerable empirical research on the relative deprivation and isolation of certain groups of older people in rural areas (Laws and Harper (1992) provide a summary of much of this in the USA and UK), I am here more concerned with the interactive relationship between constructed models of ageing and similar constructs of rural places.

Taking as its stance the notion that recent Western society has negated the natural experience of the decay and death of all human life (Cole 1993), I shall examine the alternative models of late adulthood that have emerged in its stead, arguing that we can understand the political, economic and social manipulations of later life only if we acknowledge as central the role of the body and our cultural conceptions and denials enveloping the mind–body split. In particular, I shall argue that finality – an acceptance of the decay of the *physical body*, has been replaced in our culture by the notion of frailty – a transitory state that can be overcome by modern science. The true idea of the physical body is thus replaced by the *cultural body*, which aims to deny its physical reality, a denial reinforced by yearning for the *metaphorical body* which has never been, created through the construct of nostalgia. This cultural confusion over, or denial of, the integrity of human existence leads to the marginality of those in later life, through society's behaviour towards the *social body*. While all four constructs have relevance for those residing in rural areas, as they do for all older people residing in all environments, two stand out in particular as

181

being of specific interest to those working with the rural. The metaphorical body, with its emphasis on the nostalgic, employs many of the frames used in the sentimental construction of the rural idyll, and indeed in the construction of later life the theme of the rural is often evoked. Similarly, the construct of the social body has specific relevance for the reciprocal construction of rural places and peoples. It is this that the final section of the chapter will highlight: the relationship between action, identity and place, in relation to the in-migration to rural areas of older people following retirement.

FINALITY: THE NOTION OF THE PHYSICAL BODY

As both Achenbaum (1979) and Cole (1993) have vividly shown, historical and cultural contexts have critically impacted upon the societal and individual experience of later life. Tracing the North American understanding of ageing, Cole argues that early American settlers, inspired by their faith and vision of life as a spiritual journey, found strength and personal development in the acceptance of decline and decay in old age. Hope and triumph were linked dialectically to tragedy and death. However, this acceptance of the human condition was swept away as the

> rise of liberal individualism and of a moral code relying heavily on physical self-control marked the end of the American culture's ability to hold opposites in creative tension, to accept the ambiguity, contingency, intractability, and unmanageability of human life.
>
> (Cole 1986: 122)

Initiated in North America by the Protestant ante-bellum revivalists, a similar code of industry, self-denial and restraint can be identified in the British Victorian male, swept forth by liberal capitalism. Thus the former existential integrity of the progress of life

> was virtually lost in a liberal culture that found it necessary to separate strength and frailty, growth and decay, hope and death. A society overwhelmingly committed to material progress and the conquest of death abandoned many of the spiritual resources needed to redeem human finitude.
>
> (Cole 1986: 123)

Driven by their belief in the power of the individual will, Victorian moralists rationalised experience in order to control it. Ideological and psychological pressures to master rather than accept old age generated a dichotomous tension in the perception of later life, one that retains power, albeit with different categories, today.

Rather than acknowledge ambiguity and contingency in aging, Victorians split old age into sin, decay and dependence on the one hand; virtue, self-reliance and health on the other ... anyone who lived a life of hard work, faith and self-discipline could preserve health and independence into a ripe old age; only the shiftless, faithless, and promiscuous were doomed to premature death or a miserable old age.

(Cole 1986: 123)

Thus acceptance of finality as a natural part of the human condition, which could enrich and inform all life, was slowly replaced throughout the later nineteenth century by a view that physical decline was 'one of life's problems to be solved through willpower, aided by science, technology, and expertise' (Cole 1993: xxii).

By the middle of twentieth century the *personal* responsibility part of the continuum had been almost completely replaced by *scientific* responsibility. Modern science had clearly documented the physical decline in later life. But rather than there being a return to an acceptance of this decline, and the rebuilding of an ethos which fully acknowledged it to be a part of the human experience, modern science, having discovered the 'problem' of old age, resolved to solve it. 'Unable to infuse decay, dependence, and death with moral and spiritual significance, our culture dreams of abolishing biological aging' (Cole 1986: 129).

Within Anglo-American society (and subsequently across many other cultures (Harper 1997)), acceptance of finality is thus replaced with the cultural construct of human frailty: a notion that death and decline can be reversed, cured, somehow transcended by science, based around cultural constructs of how the body should be rather than around human experience. Thus rather than see infirmity as a common experience of all human life, but one that is expanded in later life, thus providing an opportunity to address infirmity at all ages, we see the frailty of older people as setting them apart from younger persons. And in this distancing not only does ineffective communication between the generations occur (Giles and Coupland 1991), but all those defined as old (as earlier mentioned, currently a chronological or economic definition of 65 or post-retirement is used) are also contaminated with the same construct.

FRAILTY: THE NOTION OF THE CULTURAL BODY

The cultural construct *frailty* clearly reflects the position of contemporary Western rationalist thought, which continues to regard the body as an adversary to the self, the latter being more purely reflected by the mind (Frank 1990, 1991; Turner 1984, 1991). It is only by retaining the body as central to this discussion, and exploring the contradictory location of the body – all bodies, not just ageing bodies – within our current conceptual framework that we can begin to explore and understand this confusion over ageing. Indeed, as feminist and postmodernist writers are increasingly pointing out, this continual conflict within Western rational thought between objectivity and subjectivity, body and self/mind/spirit, is at the very fundament of the crisis in Western philosophy. The current 'crisis of reason' is

> a consequence of the historical privileging of the purely conceptual or mental over the corporeal; that is, it is a consequence of the inability of Western knowledges to conceive their own processes of (material) production, processes that simultaneously rely on and disavow the role of the body
>
> (Grosz 1993: 187)

From a rationalist perspective, therefore, the human condition is intelligible. In rationalist metaphysics body and spirit divide: the spirit is pure and enduring, the body irrational and transient. So long as bodily concerns do not disrupt rational spiritual life, the finality of the body can be ignored, but when the ageing body interrupts the progress of the spirit, the body as object overwhelms the self at its very centre. As Gadow (1986) has argued, when experience is dominated by the body's apparent dysfunction, the self repudiates the body in order to escape being contaminated by its deterioration. This tension between subject and object, self and body, thus becomes a lived contradiction cultivated by the self in order to avoid becoming one with the frail and failing body,

> but the dignity purchased by disowning the frail body is ultimately self-defeating, for the self thereby surrenders its most essential freedom, that of deciding how it shall regard itself ... no longer does the contradiction lie between self and body; now the contradiction exists within a self, a subject that regards itself as an object.
>
> (Gadow 1986: 239)

Thus, under rationalist thought, the integrity of the self–body relation is destroyed, and in distancing itself from the body's failure the self makes of itself an object. Furthermore, not only is the possibility of unity between the ageing self and body denied, but so too is the relationship between the self and society. Thus, not only does frailty replace finality for the individual, but it reflects back to reinforce this construct for the rest of society.

As was earlier mentioned, this denial of death and decay by modern Western society has recently attracted much debate within gerontology. In this context, not only has the analytical construct frailty much to offer, but furthermore I shall argue that we can more fully understand its construction by reference to some of the recent feminist writings on the body (Bordo 1989; Dallery 1989; Gatens 1991; Wilshire 1989; Young 1990), in particular that on the creation of sexed knowledge (Grosz 1993, 1994). In her recent text *Volatile Bodies* (1994), Grosz, herself influenced by the writings of the French feminists Irigaray and Cixous, clearly demonstrates the relationship between rational thought and the body–mind split, and between rational thought, disembodied knowledge and maleness. In particular, she argues that knowledge[1] is itself sexed, and within contemporary thought is male. However, she believes that the masculinity or maleness of knowledge remains unrecognised as such because there is no *other knowledge* with which it can be contrasted.

> Men take on the roles of neutral knowers, thinkers and producers of thoughts, concepts or ideas only because they have evacuated their own specific forms of corporeality and repressed all traces of their *sexual* specificity from the knowledges they reproduce. . . . Many features of contemporary knowledges . . . can be linked to man's *disembodiment*, his detachment from this manliness in producing knowledge or truth.
>
> (Grosz 1993: 204, 205)

Furthermore, men have been able to rationalise their domination of the production of knowledge by claiming that their interests are universal and sexually neutral. This has been possible through the constructed correlation of men with the category mind and women with body, the latter taking on the function of representing the natural. 'By positioning women as the *body*, [men] can project themselves and their products as *disembodied*, pure, and uncontaminated' (Grosz 1993: 209). This is not the result of a male conspiracy to create knowledges in their own image, but rather the result of the social meanings, values and knowledges marked upon the male body from infancy. The resulting construction is that women's bodies are constructed under this system of knowledge as castrated, lacking and incomplete. However, drawing on the work of Irigaray (1985), Grosz (1993: 208) argues that our understanding of this system of knowledge is not limited to representations of women but rather 'must include the elision of any *maleness* or masculinity in the perspectives and enunciative positions constitutive of knowledge'.

The work of both Irigaray and Grosz thus demands that we at least recognise the sexualisation of all knowledges, acknowledge the sexually particular positions from which knowledges emanate and by which they are interpreted and used. This insight is clearly not confined to gender discourse; there are clear statements of relevance for other problematic and contested relationships, including that of age.

A more complete understanding of the ageing body and notions of control can be attained if we acknowledge that control under the dominant system of knowledge can be directly related to the gendered experience of the body. This operates from within and without. We all have a degree of control over our own bodies by the virtue of inhabiting them; yet through various discourses of power, others also control our bodies for and on behalf of us. Such disciplinary regimes 'normalise' the body (Foucault 1979). As Douglas (1966) long ago pointed out, sexually differentiated biological processes and the bodily emissions concerned with these processes are signified in all cultures. However, clearly the element of control that men and women have over these emissions is different and significant. For the experience of control is learned in the first place from the body, and the experience of living within a body. Thus while male seminal emissions are under the control of the male – he can generally determine when he wishes them to occur – the varied and complex emissions of a woman – menstruation, lactation and even childbirth itself – are not under her direct control. She cannot determine and control the timing of the flow.

This is not necessarily to imply biological determinism or essentialism. That men and women experience their bodies differently has long been recognised (Sawaki 1991), (though not always acknowledged). The key point is that the *interpretation* of these experiences is socially *constructed*. Indeed, while there has been an extensive debate over sex and gender between the essentialists and social constructionists, Schilling (1993) has recently pointed out that bodies are simultaneously constructed and real; the distinction between and acknowledgement of sex and gender is required by both camps. Indeed, Grosz (1993, 1994) herself argues that sexual differences, like class and race differences, are bodily differences but these are not immutable or biologically pre-ordained; rather, they are experienced through the medium of the body.

However, as all bodies age, control as defined through male experience is reduced: incontinence, weeping orifices, bedsores, lack of control over limbs occur. Thus as male bodies age so this essential control over bodily emission starts to fail. Urinary incontinence is widespread among men in late old age owing to the enlargement of the prostate; impotence spreads (Evans 1991). Male representations of female bodies as leaking and draining has been defined as the key cultural concept behind patriarchal control (Kristeva 1982). These same bodily representations are used to construct the models of later life, perpetuating the degradation of loss of bodily control with old age.

This is all based on one system of control as defined in order to allow the male relationship with his body to be one *of* control. Yet control, like power, is just a matter of situated perception and definition. It can thus be argued that if absolute control of the body, defined through male experience, were not the overarching notion of adulthood, the natural lack of control and associated *frailty* of extreme later life would not be so stigmatised.

Drawing on the work of Foucault (often critically), Sawaki (1991) has highlighted the relationship between knowledge, power, and the increasing medicalisation of disciplinary normalisation. Not only do definitions of mental and physical health become integrated into other disciplinary forces, but those such as advanced medical technologies operate to control the body through techniques that simultaneously render it more useful and more docile. In this manner new norms of later life based on a medical solution to physical and mental decay are constructed. Modern medical technologies thus define, reproduce and discipline both the construction and the experience of old age. In this we see the continuous attempt to return control to a neutral/male body. This is not a new process, as Cole (1993) has pointed out in relation to the emergence of the 'scientific management of ageing' in nineteenth-century America. Though this was seemingly gender neutral,

> the search for normal aging had a decidedly masculine flavor. Working in their new research laboratories, male scientists searched for the cause of aging and the means of preventing senility. These men were often aging themselves, and personal concern about declining influence, sexual potency, and productivity is sometimes evident in their writings. They sought a 'normal' old age that contained an unstated ideal of health or maximum functioning – the 'good' old age of Victorian morality.
>
> (Cole 1993: 199–200)

Of particular relevance here is the notion of embodied and disembodied knowledge, of male and female sexed knowledge, and the implications of this for our understanding of the construction of later life. As feminists have long contended, under patriarchal knowledge systems women are always embodied, men are not (Grosz 1994). Yet I suggest that this pertains only while men remain *not* the 'other' – for example, not black, not gay, not disabled and not old. *Men become embodied as they age.* In the same way that queers[2] are defined by sexuality and thus have bodies, straights do not; blacks are defined by the colour of the skin and thus have bodies, whites do not; so old people are defined by the ageing body, young people are not. Thus as disembodied men age so they gain bodies, and become embodied. In later life, therefore, men, like women, are forced to recognise the 'other' as a defining force in their own construction and experience. Re-read in the frame of sexed knowledge, what these ageing scientists were seeking was to retain disembodied control for themselves and for their fellow *men*. In so doing they re-created the *physical body* within their own sexed cultural contexts and produced a *cultural body* over which they could retain control.

NOSTALGIA: THE NOTION OF THE METAPHORICAL BODY

A partial resolution of our denial of the physical body and of our finality occurs through the symbolic appropriation of the nostalgic; we turn to metaphors of how old people should be. Smith (1992), for example, suggests that the poet Robert Penn Warren was able and eager to transcend the physical decline of his spent grandfather in the poem 'Court Martial' when he heard the tale of his involvement in the American Civil War:

> he did not see his grandfather as rugged until after he heard him confess that he had hanged ambushers while serving as a cavalry captain during the Civil War. Until that moment his grandfather had seemed dependent, passive, and unable to be useful because he was 'shrunken gray' with 'Tendons long gone crank and wry' . . . Immediately after hearing his grandfather's confession, Warren . . . imagined his grandfather as a strong horseman riding alone and 'large in the sky' with 'No speculation in eye'.
>
> (Smith 1992: 222)

Work on representations of older people in art and photography (Blaikie 1994; Blaikie and Hepworth 1997; Covey 1991) also provides vivid examples of the nostalgic. Blaikie and Hepworth's (1997) analysis of images of old age within sentimental Victorian art highlights the association of grandmotherhood, femininity and country gardens, Corbin's (1986) 'perfumed domestic atmospheres'. Discussing Sarah Eddy's painting of the women's activist Susan Anthony on the occasion of her eightieth birthday, they describe the portrait as being suffused with a rosy glow, as the activist is transformed into a benevolent, stabilising granny, framed with the iconographs of acceptable nature in the form of posies of flowers. Indeed, the link between nostalgia for nature and that for old age is further emphasised by their assertion that the positive aura surrounding the granny figure is most persuasively represented by symbols of an idealised natural world. They contend that the most enduring images of later life that have come down to us from the Victorian period are of figures in country cottage gardens or kitchens. While this is clearly in part the wider reflection of the general urbanised Victorian longing for a lost world of nature (Wood 1988), Blaikie and Hepworth quote Gillet's (1990) view that the artists were also responding to the relentless pull of respectability in Victorian society. The nostalgic was in part reinforced by a society longing for the past as a way of stabilising the contemporary. And who better to do this than the head of them all: Queen Victoria, in white lacecap and plain black dress, the generic granny uniting in one symbolic community the highest and lowest in the land (Blaikie and Hepworth 1997).

It was not only the Victorians who looked to the past in this way.

Blaikie and Hepworth also discuss the 1994 Barbican exhibition of photographs 'All Human Life', selected from the Hulton Deutsh Collection. Out of 15 million photographs, both the catalogue and exhibition leaflet featured older people on the front page – in the first case, a couple lovingly celebrating their golden wedding anniversary, in the second a portrait of 'Old Tom Owen' having his head shaved. The justification for selecting these two images from an exhibition in which those in later life were not especially predominant was that:

> Among my own personal favourites are pictures of some of the old *characters* who you might still, if you are *fortunate* enough, *stumble across* every now and again ... people like Old Tom Owen, the *hedge-laying specialist*, loving every moment of his haircut.
> (Bernard 1994: 9, quoted in Blaikie and Hepworth; my emphasis)

As Blaikie and Hepworth comment, 'Old Tom' is a figure of yesteryear, a motif synthesising times gone by and old age in the same frame. Note also the nostalgia for the rural, vanishing skills, disappearing communities, countryside characters. Indeed, this continual linking of old age with the rural – the concept of the ever-cycling round of birth, death and return to nature – resounds throughout such images. Whether it be the Victorian obsession with the country garden as the context to the Victorian granny (Covey 1991), the commercial exploitation of the rugged, earthy, natural old to advertise healthy products (Blaikie 1994), the use of rustic imagery to promote retirement communities or pension plans (Featherstone and Hepworth 1995), in both contemporary and historical iconographs the nostalgia for the rural that never was is coupled with the nostalgia for a time of life that rarely is. As Stewart has argued, as the search for the authentic becomes critical in contemporary society, so

> experience is increasingly mediated and abstracted, the lived relation of the body to the phenomenological world is replaced by a nostalgic myth of contact and presence. 'Authentic' experience becomes both elusive and allusive as it is placed beyond the horizon of present lived experience, the beyond in which *the antique, the pastoral*, the exotic, and other fictive domains are articulated.
> (Stewart 1993: 133; my emphasis)

The prevailing motif of nostalgia is thus the denial of space between nature and culture. The 'nostalgic's utopia is prelapsarian, a genesis where lived and mediated experience are one, where the authenticity and transcendence are both present and everywhere' (Stewart 1993: 23), and in this process of distancing, 'the memory of the body is replaced by the memory of the object, a memory standing outside the self and thus presenting both a surplus and a lack of significance' (Ibid.: 133).

As Stewart continues, 'Nostalgia is a sadness without an object, a sadness

189

which creates a longing that of necessity is inauthentic because it does not take part in lived experience' (Ibid.: 23). It stands both before and beyond lived experience. Thus that which nostalgia seeks has never existed except as narrative, and in its continual absence it reproduces itself as a constant but distant longing. In our cultural confusion over old age, the narrative of later life, the *metaphorical* or symbolic body is preferred to the realities of the norms of human ageing – as is the narrative of the rural idyll. As Lowenthal (1986) reminds us, that which is weathered or decayed, or bears the marks of long-continued use, looks aged and thus seems to stem from the past. The nostalgic connection that old people maintain with a past before our time remains a more comfortable metaphor within which to accept the human condition than the reality of decline and death as a mainstream experience.

Clearly, there are a range of images of later life, and one must be careful not to confuse the nostalgic with attempts over the past twenty years to promote positive images of ageing. Yet at one level this only augments the notion of old age as a chronological, biological construct primarily defined for political and economic reasons, rather than face up to ageing as a continuous process that affects individuals in different ways and has a definite end in mind: decay and death.

That such images arose in reaction to the burden literature is apparent in Achenbaum's (1995) reflexive essay on the visual exhibition 'Images of Old Age in America, 1790 to the Present' selected by himself and Kusnerz in 1976. The exhibition was transported by Smithsonian Institution Traveling Exhibition Services to locations across the USA and was prominent at the 1981 White House Conference on Aging. In retrospect, Achenbaum notes the dearth in the selection of images of ones that emphasised the risks of dependency or death in later life. Thus the older Americans in the first section (1790–1864) were remarkably ebullient, regardless of their social conditions or economic status; in the second section (1865–1934) there were no pictures of the very old;[3] and in the contemporary section the old-old were reflected through the use of photographs which we might well describe as portraying 'characters': a Texas Mexican husband and wife; Mennonite farmers; a Hispanic women; a male ex-slave in Georgia; and a Spanish-American farmer. Only two were of very old people – one an ex-slave. Was it, Achenbaum asks, that we were so influenced by the notion of positive ageing that we subconsciously sought to portray only this? Was it that there were very few images available to us that reflected the darker side of later life? Or was it that we did not want to face the reality of human decline? Within our frame the question would be posed: was there not a certain *nostalgia* about the images selected – a yearning for the connection that the old maintain with the past, a turning to the metaphorical body as the more comfortable symbol of the human condition?

For as with the rural (Short 1992), are we not still affected by the Victorian longing for a perfumed, rosy later life? Indeed, as Hepworth and Featherstone (1995) discuss in relation to the soft-sell of consumer goods, services and lifestyle to the over-fifties, this exploits the hopes and aspirations of many older people, in particular the dream of an idyllic retirement, which is a persistent legacy from our Victorian past. Yet this dominant consumer culture image of a positive old age not only confounds the idea of later life as being one homogeneous experience, but also leaves those in late life, in particular the third-agers or young-old, on an extended plateau of active middle age typified in the imagery of positive ageing as a period of youthfulness and active consumer lifestyles. This continues to reinforce 'the de-meaning of age rooted in modern culture's relentless hostility toward decay and dependency ... later life floats in a cultural limbo' (Cole 1992: xxvi); continues to deny the *physical* body; the *finality* of human life, and to find comfort in the mass marketing of *nostalgic* images of later life; and the *metaphoric* body. Only when we are able to accept the changing and declining body as part of the normal experience of humanity, one that is crucial to the defining of human life, and (while recognising the extreme diversity of the human ageing experience), return it spatially and metaphorically to centre-stage within human experience, will the inclusion of all older adults within the mainstream occur. Yet this requires recognition that all experiences are valid, not just those that currently dominate twentieth century discourse.

MARGINALITY: THE NOTION OF THE SOCIAL BODY

The marginality of elderly people is clearly multi-dimensional: they are marginalised spatially through residential segregation, politically and economically through retirement from positions of power and formal employment, socially through intergenerational segregation, and symbolically with the aged forming a distinct symbolic group. Yet while a variety of structural, cultural and indeed psychological explanations have been produced to account for this marginality – political economy (Walker 1981 in the UK; Estes 1979, 1986 for the USA); structured dependency (Townsend 1981); intergenerational competition (Bengtson and Kuypers 1985); disengagement (Cunningham and Henry 1961) – the context to all these is that the perceived frailty of the older adults past a certain chronological age provides the justification for certain acts of action towards them – be these at the individual or societal level, structural or behavioural.

Fundamental to this marginalisation are thus physiological changes that naturally occur within older bodies, yet are defined by existing, Western, patriarchally based knowledge systems as being somehow outside normal human experience. These physiological changes are thus seen as making older people unsuitable for mainstream activities, and these messages are further internalised by older people, who learn to accept this constructed self-image. Thus abrupt *mass retirement* at a set chronological age as the normal experience of those in later life is a very recent historic experience even in the West, and one that is as much politically, economically and culturally constructed as emerging from any physiological or mental requirement of older people themselves (Harper and Thane 1989). Similarly, there is no clear evidence that intergenerational residential segregation is beneficial to any age. And there is a growing body of academics and medical personnel who are speaking out against the allotment of older people to specialised geriatric medical services, rather than retaining them within adult provision *per se*. This is not to argue that as adults age, so there do not arise physiological changes which may react well to certain medical treatments (Brody 1995), or that certain behavioural changes may not occur as result of these psychological developments (Giles and Coupland 1991), but rather that these need not exclude all older adults (i.e. those over 65) from activities and services available to the main population. Instead, they should be seen as but a part of constant change throughout adult life, a continuation of the spectrum of human development.

It is thus clear that chronological, biologically based age deterioration has been transformed by contemporary social structures into social and cultural signs, which are manifested in complex social relations. These relations, however, are located in particular temporal and spatial contexts, and are constantly contested and re-created in struggles over identity politics. As Foucault (1983) reminded us, such identities are constructed in discourses and thereby constituted in discursive realms. All knowledge is situated in people, places and times. It is thus of interest to explore the manner in which these identities are played out in rural areas; to question whether there is something specific to the rural that makes the experience of being an older adult living within the countryside unique to that particular location.

As was earlier stated, all four constructs have relevance for those residing in rural areas, as they do for all older people residing in all environments. However, the notion of the *metaphorical body*, with its symbolic appropriation of the nostalgic, not only employs many of the frames used in the construction of the rural idyll, but also evokes these rural metaphors to construct images of later life. More specifically, the notion of *marginality* in the construct of the *social body* can be seen to have specific relevance for rural places and peoples.

ACTION AND THE IDENTITY OF PLACE

The perceived and actual spatial isolation of older people within rural communities emphasises notions of the other, beyond. Thus, to call on Foucault, despite the fact that those in later life are defined by a variety of representations, it is clear that most of the relations between those over and under 65 are primarily constructed through difference. This construction of difference not only supports unequal power relations between the young and the old, but also creates a homogeneity that excludes the complexities of difference. Thus the other, the one of difference, becomes located in opposition to the one who is not the other, allowing the latter the power to construct the other in their own terms, and the identification of the other allows for the production of identity, and normalises the relationship of control. Whether isolated in rural areas by rural out-migration of a younger population, or self-selecting of these areas through retirement in-migration, the fact that this older population is seen to be the other, metaphorically outside the centre, is compounded in those areas where older people are not only metaphorically at a distance, but also *perceived* to be spatially thus.[4]

I shall conclude this chapter by taking one aspect of this marginality and suggesting that residential mobility of some older persons in the Western world in late age not only helps in the construction of this period of later life, and in particular in the marginality of that later life, but also constructs particular places, and, in particular, marginal places. Thus in moving from the urban to the rural on leaving economic employment, men and women do not only change their own status, they change the status of the rural. On the one hand the *rite de passage* of retirement from economic employment is intensified by a spatial relocation. In spatial migration those retiring accentuate the transition. The spatial shift from centre to margin thus not only represents a shift in activity spaces for elderly people, but also reinforces the transformation of their own personal identities as they chronologically move across the metaphorical space of retirement into old age. They are thus transforming their identity through a move from the centre to the margin both spatially and metaphorically. However, it is not only the identity of older people themselves that is affected. I believe we can argue that such spatial relocation by a population constructed as being marginal further confirms the marginal identity of the spaces into which they are moving. Thus the rural becomes more marginal in association with its population, and the elderly become more marginal in association with their residential space.

CONCLUSION

Rather than take either the rural or the elderly as a self-evident category, this chapter has attempted to show how attributes associated with later life are constructed and used in different ways and contexts. It has argued that apparently natural attributes associated with later life (decay, incontinence, outside, beyond, embodied) have been set up in opposition to youthful attributes, and in particular youthful male attributes (growth, control, inside, within, disembodied) in order to define and control what is in effect a specific economic group, 'retired' people, whose natural association with a chronological, biologically based age category is arbitrary, and whose homogeneity does not exist in terms of gender, ethnic group, class, sexuality nor life course experience. I have argued that these representations have significance for all those excluded from the centre by association with contemporary representations, but can be seen to have a specific significance for those elderly people residing in rural places, and for these places themselves, through the reciprocal association with the margin.

NOTES

1 We are here discussing the traditional knowledge of power: those of the sciences, humanities, social sciences, law and medicine. Grosz (1993) confines discussion to knowledge of the social sciences and humanities; Irigaray (1985) has extended this to include other traditional knowledges.
2 'Queers' here refers to people other then straight heterosexuals and thus includes gays, lesbians, bisexuals and transvestites, as well as heterosexuals engaged in alternative practices. See Berlant and Freeman (1993).
3 Achenbaum distinguishes between 'young-old' or third-agers, 'old-old' and 'very old' on a spectrum of increasing age.
4 This is not to argue that the countryside *is* marginal in relation to a preconceived centre, but to suggest that in late-twentieth-century Britain it is still perceived as being thus.

REFERENCES

Achenbaum, A. (1979) *Old Age in the New Land*, Baltimore: Johns Hopkins University Press.
—— (1995) 'Images of old age in America, 1790–1970', in Featherstone, M. and Wernick, A. (eds) *Images of Aging*, Routledge, London, 22–47.
Bengtson, V. and Kuypers, J. (1985) 'The family support cycle', in Munnichis, J., Mussen, P. Olbrich, E. and Coleman, P. (eds) *Lifespan and Change in Gerontological Perspective*, Academic Press, California, 257–74.
Berlant, L. and Freeman, E. (1993) 'Queer Nationality', in Warner, M (ed.) *Fear of a Queer Planet*, Universtity of Minnesota Press, Minneapolis, 193–229.
Bernard, B. (1994) *All Human Life*, Barbican Art Gallery, London.
Blaikie, A. (1994) 'Photographic memory, ageing and the life course', *Ageing and Society* 14.
Blaikie, A. and Hepworth, M. (1997) 'Is the medium the message? Representations of age in paintings and photographs', in Jamieson, A., Harper, S. and Victor, C. (eds) *Approaches to Ageing and Later Life*, Open University Press, Buckingham.
Bordo, S. (1989) 'The body and the reproduction of femininity: a feminist appropriation of Foucault,' in Jaggar, A. and Bordo, S. (eds) *Gender/Body/Knowledge*, Rutgers University Press, New Brunswick, 13–33.
Brody, J. (1995) 'Postponement as prevention in aging', in Butler, R. and Brody, J. (eds) *Delaying the Onset of Late-Life Dysfunction*, Springer, New York.

Cixous, H. (1980) 'The laugh of the Medusa', in Marks, E. and de Courtivron, I. (eds) *New French Feminisms*, Amherst, University of Massachusetts Press, 24–64.

Cole, T. (1986) 'The enlightened view of aging', in Cole T. and Gadow, S. (eds) *What Does It Mean to Grow Old?*, Duke University Press, Durham, NC.

—— (1993) *The Journey of Life*, Cambridge University Press, Cambridge.

Corbin, A. (1986) *The Foul and the Fragrant: Odour and the French Social Imagination*, Leamington Spa, Berg.

Covey, H. (1991) *Images of Older People in Western Art and Society*, Praeger, New York and London.

Cunningham, G. and Henry, W. (1961) *Growing Old: The Process of Disengagement*, Basic Books, New York.

Dallery, A. (1989) 'The politics of writing the body: *ecriture féminine*', in Jaggar, A. and Bordo, S. (eds) *Gender/Body/Knowledge*, Rutgers University Press, New Brunswick, 52–67.

Douglas, M. (1966) *Purity and Danger*, Routledge and Kegan Paul, London.

Dreher, B. (1987) *Communication Skills for Working with Elders*, Springer, New York.

Elias, N. (1978) *The Civilising Process*, Blackwell, Oxford.

Estes, C. (1979) *The ageing Enterprise: A Critical Examination of the Social Policies and Services for the Elderly*, Jossey-Bass, San Francisco.

—— (1986) 'The politics of ageing in America', *Ageing and Society* 6: 121–34.

Estes, C. and Swan, J. (1993) *The Long Term Care Crisis*, Sage, Newbury Park.

Evans, R. (1991) 'Advanced medical technology and elderly people', in Binstock, R. and Post, S. (eds) *Too Old for Health Care*, John Hopkins University Press, Baltimore.

Featherstone, M. and Hepworth, M. (1995) 'Images of positive ageing', in Featherstone, M. and Wernick, A. (eds) *Images of Aging*, Routledge, London, 29–47.

Foucault, M, (1979) *Dicipline and Punish*, Vintage Books, New York.

—— (1983) 'The subject and the power', in Drefuss, H. and Rabinow, P. (eds) *Michel Foucault: Beyond Structuralism and Hermeneutics*, University of Chicago Press, Chicago.

Frank, A (1990) 'Bringing bodies back in: a decade review', *Theory, Culture and Society* 7(1): 131–62.

—— (1991), 'For a sociology of the body: an analytical review', in Featherstone, M., Hepworth, M. and Turner, B. (eds) *The Body: Social Process and Cultural Theory*, Sage, London, 36–102.

Gadow, S. (1986) 'Frailty and strength: the dialetic of aging', in Cole, T. and Gadow, S. (eds) *What Does It Mean to Grow Old?*, Duke University Press, Durham, NC.

Gatens, M. (1991), 'Corporal representation in/and the body politic', in Diprose, R. and Ferrell, R. (eds) *Cartographies: Poststructuralism and the Mapping of Bodies and Spaces*, Allen and Unwin, Sydney, 79–87.

Giles, and Coupland (1991) *Language: Contexts and Consequences*, Open University Press, Milton Keynes, 371–89.

Gillet, P. (1990) *The Victorian Painters' World*, Alan Sutton, Gloucester.

Grosz, E. (1993) 'Bodies and knowledges: feminism and the crisis of reason', in Alcoff, L. and Potter, E. (eds) *Feminist Epistemologies*, Routledge, New York and London.

—— (1994) *Volatile Bodies*, Indiana University Press, Bloomington and Indianopolis.

Harper, S (1997) *Ageing Societies*, Arnold, London.

Harper, S. and Thane, P. (1989) 'The consolidation of old age as a phase of life,

1945–65', in Jeffreys, M. (ed.) *Growing Old in the Twentieth Century*, Routledge, London; 43–61.

Hazan, H. (1994) *Old Age: constructions and deconstructions*, Cambridge University Press, Cambridge.

Irigaray, L. (1985) *The Sex Which Is Not One*, Cornell University Press, Ithaca.

Jones, H. (1990) *Population Geography*, Paul Chapman, London.

Kristeva, J. (1982) *Powers of Horror: An Essay on Abjection*, Columbia University Press, New York.

Laws, G. and Harper, S. (1992) 'Ageing in rural society', in Bowler, I. and Mellis, D. (eds) *Rural Restructuring*, vol. 2. CAP.

Lowenthal, D. (1986) *The Past Is a Foreign Country*, Oxford University Press, New York.

OPCS (1995) 1991 Census General Report, HMSO, London.

Mutchler, J. and Burr, J. (1991) 'A longitudinal analysis of household and non-household living arrangements in later life', *Demography* 28(3): 375–90.

Sawaki, J. (1991) *Disciplining Foucault*, Routledge, New York and London.

Schilling, C. (1993) *The Body and Social Theory*, Sage, London.

Short, B. (1992) 'Images and realities in the English rural community', in Short, B. (ed.) *The English Rural Community*, Cambridge University Press, Cambridge.

Smith, C. (1992) 'Images of aging in American poetry, 1925–85', in Cole, T., Van Tussel, D. and Kastenbaum, R. (eds) *Handbook of the Humanities and Aging*, Springer, New York.

Stewart, S. (1993) *On Longing*, Duke University Press, Durham, NC.

Townsend, P. (1981) 'The structural dependency of the elderly', *Ageing and Society* 1: 5–28.

Turner, B. (1984) *The Body and Society*, Oxford, Blackwell.

—— (1991) 'Recent developments in the theory of the body', in Featherstone, M., Hepworth, M. and Turner, B. (eds) *The Body: Social Process and Cultural Theory*, Sage, London, 1–35.

Walker, A. (1981) 'Towards a political economy of old age', *Ageing and Society* 1: 73–94.

Warnes, A. (1991) 'The changing elderly population: aspects of diversity', *Reviews in Clinical Gerontology* 1: 185–94.

Warnes, A. and Ford, R. (1993) 'The changing distribution of elderly people: Great Britain 1981–91'. Occasional Paper 37. Department of Geography, King's College London.

Wilshire, D. (1989) 'The uses of myth, image and the female body in re-visioning knowledge', in Jaggar, A. and Bordo, S. (eds) *Gender/Body/Knowledge*, Rutgers University Press, New Brunswick, 92–114.

Wood, C. (1988) *Paradise Lost: Paintings of English Country Life and Landscape 1850–1914*, Barrie & Jenkins, London.

Young, I. (1990) *Throwing Like a Girl and Other Essays in Feminist Philosophy and Social Theory*, Indiana University Press, Bloomington.

11

ETHNICITY AND THE RURAL ENVIRONMENT[1]

Julian Agyeman and Rachel Spooner

The countryside is popularly perceived as a 'white landscape' (Agyeman 1989a), predominantly inhabited by white people, hiding both the growing living presence and the increasing recreational participation of people of colour.[2] Thus in the language of 'white' England, ethnicity is rarely an issue associated with the countryside. Its whiteness is blinding to its presence in any other form than 'non-white'.

Just as many would argue that there is no need for anti-racist or multicultural education in rural schools,[3] a common response to the idea that issues of 'ethnicity' are important in rural areas is one of amusement and derision, accompanied by the comment that 'there is no problem here' (Derbyshire 1994; Jay 1992). Yet if those defining the agenda in rural areas do not recognise 'race' as an issue, there is little chance of change being brought about. As Bonnett argues, 'we need to think beyond the deeply disturbing cliché that "White" areas do not have a "race problem" ' (1993: 175).

The study of 'race' and racism in geography, alongside the focus of anti-racist organisations, may have reproduced this omission by primarily focusing on towns and cities, those places with relatively large populations of colour (Anderson 1988; Jackson 1987; Smith 1989).[4] There is a small, but significant, population of colour living in rural areas (Jay 1992), yet these people's experiences are neglected in studies which focus on the role of rurality as a signifier of an exclusive and white national identity. Representations of the countryside are controlled by white (generally male and middle-class) people. They construct images reflecting a concern with the reproduction of a mythical and nostalgic white heritage (Agyeman 1993; Murdoch and Day 1995). Others, who are potentially threatening to the integrity of this representation, are excluded. In this way 'the rural' becomes a culturally contested and sanitised landscape.

In this chapter, we seek to illustrate that racist attitudes do exist in and about the countryside and that these affect the lives of people of colour, both as residents and as (potential) recreational visitors. We want to show how selective representations of rurality are being contested by people of

197

colour through exposing the oppressive nature of social relations vested in rurality. We begin with a look at connections between rurality and ethnicity, both historical and contemporary, drawing on some of the ideas that underpin many discussions on this subject. We then suggest how these may affect the lives and experiences of people of colour who live in rural areas, and finally we take a look at the issue of recreational access to the countryside, suggesting reasons for low levels of participation among people of colour. The Black Environmental Network[5] (BEN) has been at the forefront in encouraging an interest in the environment among people of colour. We take a look at its ideological and practical project, as well as assessing its success.

Before we begin, we feel that it is important to position ourselves as authors, coming from in some ways very different, and in some ways very similar, standpoints. We are united by our backgrounds in geography and our interests in anti-racism and the environment. Julian is a middle-class man of mixed English–African parentage.[6] He was founder, and until 1994 Chair, of BEN. Julian's role within BEN was ideological provider, a role which he combined with a deep personal commitment to BEN's early focus on issues of environment and equality. This echoes the maxim 'the personal is political', articulated by many feminist researchers (e.g. Stanley and Wise 1993), and in Julian's case led to his resignation as the political project of BEN began to drift away from his personal vision. Rachel is a white, middle-class, young woman. Here, her whiteness raises many questions concerning the power relations involved in 'cross-cultural' research, and it is important that we (and any readers) are aware that her experience as a white woman may 'constitute an active barrier to understanding other . . . experiences' (Puar 1993: 5). Moreover, Rachel's whiteness perhaps leaves her open to the criticism that she is complicit with those white organisations and individuals who, through an interest in anti-racism, participate in a process of the appropriation of the lives and experiences of people of colour. The power of representation rests firmly with white people. This is an important issue and one that we return to in terms of how BEN's radical agenda has been co-opted to a more mainstream one. We have, however, used Rachel's whiteness in this project to our advantage, particularly in terms of obtaining information from organisations that may feel threatened by Julian's blackness. This is a situation which itself begs many questions concerning racial identities.

We make these statements at the risk of simply acknowledging our positions without any recognition of them in practice, as well as reproducing the narcissistic navel-gazing of many attempts at autobiography. Furthermore, we acknowledge that this kind of theorising may not make our positions any clearer, since, as Jackson and Penrose point out, 'we may not be fully aware of how all the dimensions of our identities bear on our research' (1993: 208). However, we feel that this brief acknowledgement of our 'situated knowledges' (Haraway 1991) is important since a great deal of the work on ethnicity and rurality, while focusing on people of colour, is controlled and written by white people.

A WHITE SPACE?

Perhaps ethnicity remains largely unproblematised within notions of rurality owing to the invisibility of whiteness as an ethnic signifier. White people are beginning to understand that their ethnicity is not an external issue, but shapes the experiences of the lives of both white people and people of colour (see Dyer 1988; hooks 1992; Frankenberg 1993a, 1993b; Hall 1992a, 1992b; Morrison 1992; Ware 1992; Wong 1994). The invisibility of the dominant Self has worked through differential relations of power, allowing whiteness to remain outside the process of definition (Jackson and Penrose 1993). The ethnic Other has been constantly redefined and renamed, reinforcing its difference and marginality from a white 'norm'.[7] hooks (1991) returns the dominant gaze in her assertion that for people of colour, whiteness has never been so invisible, but signifies terror and pain.

These ethnic identities and fears are closely linked to place (Anderson 1988; Sibley 1995). Geographers have shown that space is often conflated with ideas of ethnicity and that the boundaries between ethnicities have come to offer conducive conditions for the construction of distorted cultural representations (Duncan and Ley 1993). To use a frequently cited example, for white people the 'inner city' has become a coded term for the imagined deviance of people of colour, a place of fear and a place to avoid. These ethnic identities and fears may also be mapped onto the contemporary British countryside. For white people, 'ethnicity' is seen as being 'out of place' in the countryside (Agyeman 1989a), reflecting the Otherness of people of colour. In the white imagination people of colour are confined to towns and cities, representing an urban, 'alien' environment, and the white landscape of rurality is aligned with 'nativeness' and the absence of evil or danger. The ethnic associations of the countryside are naturalised as an absence intruded upon by people of colour. Lowe notes that

> from the turn of the century period comes an aesthetic and spiritual identity with the wild, strong anti-urban and anti-industrial sentiments, and a sense of stewardship, associated on the one hand with an appreciation of the web of life and its fragile balance, and on the other hand with a patriotic attachment to the indigenous flora and fauna.
>
> (Lowe 1983: 349)

The implication is that the 'fragile balance' of social and ecological relations in the English countryside will be disturbed, perhaps irrevocably, by alien invasions, both human and non-human. This attachment to 'native' plants and its association with racist discourse potentially acts as a barrier to involvement in the countryside and will be discussed in greater detail below.

For people of colour, 'ethnicity', in terms of a white Other, signifies fear of the countryside, of white spaces. This fear becomes such that they are effectively disenfranchised from this 'unsafe' environment (Agyeman 1995; Kinsman 1993). This sense of fear, of 'dread' (Pollard 1989), has evoked a powerful response from Ingrid Pollard, a black[8] female photographer, who in her series Pastoral Interludes seeks to challenge her personal sense of exclusion from the countryside, and the idea that it is not part of the experience of people of colour in Britain. The caption to one of these photographs reads, 'feeling I don't belong, walks through leafy glades with a baseball bat by my side' (Pollard 1989: 43). Here she confronts idealised representations of British rurality with 'the reality of the Black experience' (Tawadros, quoted in Pollard 1989: 41). Although her work has achieved a high profile (Kinsman 1993; Pollard 1993), being immediately accessible, she is not alone in exposing the oppressive nature of social relations located within ideas of rurality. Others, such as Dabydeen (1992) and Agyeman (1989a, 1991, 1995), have met through their differing, yet complementary, critiques of the English countryside, using literature and ecology/environmental management respectively.

Ethnicity and rurality: some historical connections

Historically there are long-established links between ideas of rurality, ethnicity and ethnic purity. These ideas 'arose from the same (nineteenth-century) European milieu' (Banton 1977: 3), as a method of interpreting the new and emerging social relations produced by the exigencies of colonialism, and were inscribed within the urban–rural dichotomy (Bunce 1994; Lowe 1983; Lowe et al., unpublished). The countryside, at the core of Britain's national identity, was regarded as the ideal location in which to breed a healthy and moral 'race', and country people became the 'essence of England' (Howkins 1986: 69; Taylor 1991). The purity of rural areas was juxtaposed with the pollution of urban industry and commerce, and cities were aligned with racial degeneration.[9]

The association of rural areas and racial purity has been reworked and taken on added resonance with the postmodern fragmentation of identity. The destabilisation of identity offers previously marginalised groups a series of ruptures or spaces of resistance (hooks 1991; Jones 1992; Lee 1988). Identity politics is therefore a movement predominantly associated with the left, and its attempts to decentre the reputed cultural centre (white, Western, middle-class, heterosexual, male) from a position of absolute authority. A pitfall of identity politics is its tendency towards essentialism, allowing it to be appropriated by those on the right. As Mouffe (1994) points out, with the collapse of the principal political boundary 'democracy/totalitarianism', democracy has had to redefine itself. It has done so by establishing new political frontiers, resulting in greater prominence for the views and attitudes of the extreme right, who are struggling to create a closed conception of national culture (Giroux 1993; Di Leonardo 1994).

The extreme right have seized upon a populist version of British national culture, one frequently represented by allusions to a mythical rural idyll (Mingay 1989a, 1989b, 1989c). Images of this 'green and pleasant land', free from invading aliens, have been used by fascist organisations such as the National Front (Coates 1993) and the British National Party (Jay 1992; Roberts 1992), who have targeted rural areas as some of their top recruiting grounds. The National Front believed that 'Through contact with the soil the British people were to recover their spiritual and racial identity' (Coates 1993: 15). These groups have recognised the ideological potency of the rural as a white space. This is not to suggest that all those subscribing to a 'rural idyll', or a British national culture or identity, have fascist tendencies. This is simply one reading of how an ethnically absolute national culture is inscribed in space; however, it is both powerful and frightening.

Such attitudes are extreme yet similar ones are pervasive in rural England. For example, many white 'liberals' echo these attitudes. In Norfolk, the head of one of the public services has been quoted as saying, 'some staff come to Norfolk for "quality of life" and the white complexion of the area has something to do with that quality of life' (Derbyshire 1994: 21; also see Cloke 1994: 157). The danger is that racist attitudes of the extreme right have become naturalised in the language of self-proclaimed liberals.

The significance of these opinions is perhaps highlighted by the fact that *The Archers*, a British radio soap-opera termed by Smethurst 'the centre of the ideological construction of a certain form of Englishness ... conservative, sensible and not particularly egalitarian' (in Liang 1992: 148), has begun to tackle the issue. One of the newest residents in 'Ambridge' (the fictional village where the programme is set), Usha Gupta, an Asian lawyer, has been subject to racist attacks, instigated by the British National Party, including having acid thrown in her face and slogans daubed on her house. Moreover, the actress playing the role has suffered personal harassment in the 'real' world.

The idea of national culture and its connection with the countryside, as employed by the extreme right, also works on a more popular level. A sense of ownership and belonging to the countryside, and the nation itself, is often constructed through an appeal to heritage (Kinsman 1993; Agyeman 1993). In Britain this rural heritage is portrayed as white and Anglo-Saxon and is reproduced daily in popular culture and through educational practices. Other social and cultural groupings, who potentially threaten the maintenance of the dominant social order, are excluded and in this case written out of history (see Dodd 1986; Mosse 1985). In being written out of history, people of colour are denied a similar sense of attachment to the countryside, and arguably to the nation (Kinsman 1993). The discriminatory nature of such versions of history is being challenged by an increasing number of histories of Britain which are reclaiming the marginalised contribution of Other groups to the British past (File and Power 1981; Fryer 1984; Gundara and Duffield 1992; Visram 1987). Agyeman notes the extent of the presence of people of colour in Britain:

> soldiers from North Africa used the Roman environment of the Borders. They were garrisoned on Hadrian's Wall. People from Asia were brought, often as whole villages, to Britain to work in the Yorkshire and Lancashire cotton mills. Many of our stately homes were financed, built and exotically landscaped through African–Caribbean slavery and the last Maharaja of Lahore, Duleep Singh, is buried in the church of the village of Elvedon in Suffolk. Has the presence of these, and other people been routinely celebrated in visitor attraction and interpretation facilities, or has it been quietly and unceremoniously swept under the carpet?
>
> (Agyeman 1995: 5)

These histories seek to challenge the 'view of British history as "heritage" ', reminding us instead that 'the past was not "quaint" and that the processes which created it are ongoing' (Gundara and Duffield 1992: 1). History should be viewed as a process, a constant blending and redefinition of cultural practices (Hall 1990; Gilroy 1993), yet historical and racial ideological dogma obscures a more 'multicultural' reality.

Again Pollard's photography illuminates the denial of the presence of people of colour in mainstream (white) versions of British history. A further caption, to one of her Pastoral Interlude series, reads, 'a lot of what made ENGLAND GREAT is founded on the blood of slavery, the sweat of working people... an industrial REVOLUTION without the Atlantic triangle' (Pollard 1989: 44; emphasis in original). As Agyeman (1995) suggests in the passage quoted above, many of the country houses and estates forming the cornerstone of rural Englishness were funded with money from colonial exploitation and imperial expansionism (Lowe et al., unpublished). By reclaiming these histories, people of colour can develop a sense of belonging and inclusion in relation to the countryside.

Rural living

The recognition that rural areas do face issues related to ethnicity initially received little response or interest from either environmental or 'race' agencies.[10] As these issues gained wider currency, for example in the national press (Agyeman 1989b; Coster 1991; Deedes 1992), the response was the rapid publication of a series of reports by various regional and national organisations (Jay 1992; National Alliance of Women's Organisations 1994; Derbyshire 1994; Bolton College 1993). The reports cover a wide geographical area, from the south-west to Norfolk and Bolton as well as a range of the Home Counties. Not surprisingly, all these reports found an extensive amount of racial violence, harassment, condescension and bigotry, provoked by a mixture of ignorance, the uncritical acceptance of stereotypes and a resistance to the arrival of incomers. Among service providers lip service was paid to equal opportunities, and there was much evidence of institutionalised racism. The overwhelming conclusion has been that people of colour living in rural areas face isolation from both the local community and their ethnic community, as well as problems with developing and sustaining a positive sense of identity in the face of continued racism.

These reports focus explicitly on the experiences of people of colour living in the countryside (Jay 1992; Derbyshire 1994; National Alliance of Women's Organisations 1994). They came a long time after issues of ethnicity and recreational access to the countryside had been raised by BEN (Agyeman 1989a) and others (John 1988), but they have led to a far greater recognition of the issues. For example, the National Council for Voluntary Organisations (NCVO) set up a Rural Anti-Racism Project as a direct response to Jay (1992), which has five main aims:

- to develop race equality policies and practices in rural areas;
- to develop training material on race equality with a rural focus;
- to establish a national forum with a developmental role for the project;
- to encourage the development of support mechanisms for isolated individual workers in rural areas;
- to evaluate all elements of the project and the principles established, in order to identify parallels for tackling other aspects of inequality in a rural context.

The reality and salience of these issues has been reinforced by the events surrounding Paddy Ashdown, the Liberal Democrat Party leader, on a recent fact-finding mission in his (fairly rural and very white) constituency town, Yeovil. While investigating reports of racist attacks upon Asian shopkeepers and take-away owners, he was attacked by a white man with a flick knife. One report reads, 'trained in commando warfare, the MP managed to knock the weapon away and make good his escape, mission accomplished' (McCrystal 1995: 11). Such newsworthy incidents, while highlighting the presence of rural racism, may also distort the issue. Here the threatening attack on Paddy Ashdown and the heroic way he handled himself have become the focus; the presence of racism in Yeovil is almost secondary. It downplays the pervasive feeling in rural areas that people of colour are incomers and that they do not 'belong'.

Rural cultures are often very elitist, and all newcomers to villages, both white and those of colour, can be made to feel unwelcome; yet some newcomers are made to feel more unwelcome than others. Rural areas are often considered as having distinct, close-knit cultures which are generally regarded as intolerant of difference. These have variously been described as 'communities' or systems of 'local distinctiveness' (Cohen 1982). For Cohen, these local cultures shape 'how people experience and express their difference from others, and ... how their sense of difference becomes incorporated into and informs the nature of their social organisation and process' (1982: 2). In Norfolk, racial prejudice was explained away through a claim to such distinctiveness when the managing director of one company said, 'the "natives" appear to exhibit prejudice against all incomers regardless of ethnic background' (Derbyshire 1994: 21).

In this way the boundaries of any village are policed, and systems of belonging (Cohen 1982) are established. We would suggest that ethnicity is a critical signifier of belonging, both locally and in terms of national identity. Contemporary socio-economic changes occurring in rural areas have led to the formation of new class structures (Savage *et al.* 1992; Cloke *et al.* 1995). Murdoch and Day suggest that as villages become increasingly sought-after, as sites embedded with cultural capital, and as rural residents seek to exclude 'undesirable social formations', they will become more homogeneous (1995: 7). They draw on Chambers' (1990, 1993) dichotomous versions of Englishness: 'One is Anglo-centric, frequently conservative, backward looking, located in a frozen and largely stereotyped idea of national, that is, English culture. The other is ex-centric, openended and multi-ethnic' (1990: 27). Homogeneity within rural cultures is imposed here in terms of a white, Anglo-Saxon identity, and Other social groups are actively excluded. The claim to local distinctiveness of rural areas is therefore not innocent, as suggested above, but has overt racial associations.

RURAL RECREATION

The majority of people of colour living in Britain live in urban areas. Their perceptions, experiences and thoughts about the British countryside are rarely addressed or considered. It is important in terms of countryside provision that we do not overlook these alternative sets of beliefs and values. Therefore an important reflection on issues of ethnicity and rurality is the perceptions and experiences of people of colour who live in urban areas.

Users and non-users of the countryside

The (1992) Countryside Commission document *Enjoying the country-side: Policies for People* profiles people who enjoy the countryside. It suggests that 35 per cent of the population are 'infrequent users', including those people 'over 60 years of age, equally male or female, on low income, unskilled or unemployed, living several miles from the countryside in poor housing, and dependent upon public transport. *Likely to be of ethnic minority origin*' (emphasis added). Why are people of colour infrequent recreational users of the countryside? Various explanations have been put forward, but here we review the four key factors put forward by Agyeman (1989a): cultural, economic, time and racism.

First, in terms of cultural factors, Agyeman notes that generally 'many black cultures have a folklore or mythology which reveres and respects the countryside, but does not regard it as a leisure resource. Also, the countryside is associated with backwardness, the city with progress' (1989a: 336). Thus the countryside has different meanings in different times and places, evidence that landscapes are cultural constructions. Further, conflicts may occur when meanings held by different cultural groups overlap (Bell and Valentine 1995; Cloke and Milbourne 1992). The contemporary countryside in Britain clearly means different things to different people. For example, in Anglia TV's *Countrywide*, shown in 1992, a young African–Caribbean man from Luton, on being asked about his perceptions of the countryside, described a trip with his partner to Tilbury, an industrial port in Essex, and how people stared at them. His perception of country-side, unlike that of many countryside providers, was anywhere outside his home area.

Malik (1992) develops this idea in her study of Asian attitudes towards recreational usage of the countryside. She investigated working-class inner-city white and Asian attitudes towards the countryside, and compared these to suburban middle-class Asian and white attitudes. She found that class was a determining factor in that both the white and Asian inner-city residents were less likely to use the countryside than their white and Asian suburban counterparts. However, ethnic divisions within these classes pro-duced significant differences between Asian and white people in terms of obtaining information (word of mouth was very significant for inner-city Asians) and choosing where to go (Snowdonia and the Lake District were popular with suburban Asians, who were largely Indian, because these areas reminded them of Indian hill stations).

Malik concludes that

> throughout the study, it was clearly evident that the white suburban respondents felt by far the deepest personal link with the countryside, fostered by a strong sense of heritage. Many had friends or family living there, and it is probable that for them, countryside use had

formed an intrinsic part of their cultural upbringing (for example, via family visits, tales of the past, bedtime stories etc.).

(Malik 1992: 38)

Clearly, having friends and relatives in the countryside is one key cultural factor that people of colour generally lack. Again, however, this is a generalisation that falls prey to the idea that people of colour do not live in the countryside. For example, Jay (1992) estimates that there are between 26,200 and 36,600 people who were born in the New Commonwealth and Pakistan, other parts of Africa (except South Africa), other Caribbean countries and Latin America, other Asian countries, the Middle East and Turkey, now living in the south-west peninsula (Cornwall, Devon, Somerset and Dorset). Malik also touches on the idea of heritage. We have shown how British heritage is constructed as white and rural, and may evoke patriotic responses among white residents; it may also leave people of colour feeling that they do not share the same attachment to such symbolic icons. We develop this idea further below.

Second, economic factors may be important in the case of large urban areas such as London, where 'getting out' is often difficult and costly. In other urban centres the issue of access is compounded by the disproportionate concentration of people of colour in inner cities. Other, smaller cities are very close to areas of open countryside, yet despite the minimal cost incurred this also remains largely unused. This assertion is backed up by the research undertaken in Bolton (Bolton College 1993) and Leicester (Environ 1994a, 1994b).

Third, factors of time are important in explaining why people of colour are infrequent users of the countryside. In common with other people in lower socio-economic groups, people of colour generally have less 'free time' in which to visit the countryside even if they wanted to (although, as Malik's (1992) survey showed, middle-class suburban Asians were more likely to visit rural places than inner-city white or Asian people). Free time for many people of colour is devoted to 'intra-community' activities and activities such as further and higher education in order to enhance life chances.

Racism is the fourth key factor in explaining why people of colour do not visit the countryside in numbers commensurate with their numbers in society generally (Taylor 1993). Whether the fear of racism is perceived, or real, is irrelevant. The fact is that many would-be visitors to the countryside are frightened of the potential reaction from local white people. As Malik states,

it appeared to be more the anticipation of abuse or unacceptance in the countryside, rather than any direct experiences of racism on previous visits, that deterred many people from going at all, and prevented a complete feeling of safety and relaxation for others while there.

(Malik 1992: 38)

Seen by the majority of people of colour as a 'white domain', the country-side is intimately and inextricably linked to notions of national identity (Daniels 1993; Lowenthal 1991). Indeed, the word 'country' has the double meaning of both the countryside and the nation (Bunce 1994; Short 1991; Williams 1983). Racism, in its cultural form, aligns national belonging with ethnicity, and this, in turn, is articulated 'through landscape imagery and ideology' (Kinsman 1994: 1).[11] Agyeman notes that

the countryside, changed out of all recognition as it is by the ravages of agriculture, is still portrayed by advertising and the media as the true keeper of Anglo-Saxon culture. It is seen as a timeless, unchanging place, full of the fruits of the earth, with native and natural plants and animals. It harbours a whole host of patriotic reactions in the collective psyche.

(Agyeman 1989a: 336)

The notions of 'native' and 'countryside' are seen by many as being linked (see Lowe 1983). To most conservationists, native plants represent 'the natural order'. In the popular imagination there is no place for 'the alien'. Indeed, Wright, when discussing conifers in the British landscape, notes that 'they are ... alien imports, plainly lacking the cultural credentials of the native broadleaf' and that, 'like other immigrants, these fir trees all look the same to the affronted native eye' (1992: 6). As a landscape element, plants, especially those which are native, as opposed to those which are alien, are seen as the 'rightful' inhabitants of British landscapes.

Conservationists' dislike of 'alien' plants is well documented (Barker 1994; Nicholson 1987/8), but perhaps no author has articulated the links with racism as well as Fenton. He notes that

perhaps dislike of alien species is indeed similar to racial discrim-ination – wanting to preserve the genetic integrity of one's own stock (a natural human failing). Alien species are welcome in strictly defined areas (gardens) but must not be allowed to pollute the native culture (the wider countryside).

(Fenton 1986: 21)

Agyeman comments that 'black people in the countryside are perhaps seen as "polluting" human "aliens"; they are perceived as being "out of place" ' (1989a: 336). Journalists such as Wright (1992) and Schoon (1992) have mined a vein of parochialism in attempting to popularise the debate about 'natives' and 'aliens' by appealing to xenophobic attitudes. Schoon's (1992) choice of archaic and pejorative phrases is wide-ranging and unguarded.

One can only assume that his sources fed him their own prejudices, which he then adapted into emotional populism. He talks of 'encroaching foreigners', 'running riot', 'ferocious, fast growing foreign plants', 'the villainous and the benign', 'acceptable aliens', 'staggering penetration', 'ruthlessly ousting the natives', 'pink and green Japanese terror' and plants which 'brutalise the native flora' (1992: 7). This undisguised xenophobia, including sexual metaphor ('staggering penetration'),[12] is an indication of the depth of feeling (and fear) which the issue raises.

Doughty, like Fenton (1986), takes the argument to its logical conclusion, by noting the popular comparison in the nineteenth-century USA between alien plants or animals and human immigrants. He discusses the feelings of Americans to the immigration of the English house sparrow into the USA in the nineteenth century and notes that

> sparrows and immigrants had 'low morals', reproduced at amazing rates, and appeared to be plotting and conspiring to exploit the United States at the expense of native-born Americans. In contrast, native birds were clean, tidy and hardworking who preferred country living and fulfilled the 'yeoman myth'.
>
> (Doughty 1978: 28)

He continues by noting that, according to Berrey's *American Thesaurus of Slang*, 'Irishmen were also nicknamed sparrows because they were so numerous and prolific' (ibid.). Again, the danger is that these attitudes have become naturalised in the language of mainstream white liberals.

Despite a growing interest in rural racism the issue of recreational access has not been widely addressed. The lack of opportunity experienced by people of colour in terms of countryside recreation has rarely been theorised or explained in the context of fundamental notions of racism and marginalisation. Consequently, it is suggested, any attempts to tackle rural racism have simply reproduced the status quo when it comes to recreation. The organisation BEN has been a lone voice in recognising the connection between the broader problem of racism in rural areas and the absence of people of colour from countryside recreation. The following section of this chapter will outline how, in a practical sense, BEN has attempted to confront the issue of inequality of recreational opportunity in addressing rural racism.

THE BLACK ENVIRONMENTAL NETWORK

A challenge to the denial of the presence of people of colour and the whitewashing and exclusivity of rural areas is being mounted by various groups and individuals (see, for example, Agyeman 1989a; Eaton 1994; Pollard 1989). In terms of promoting access and encouraging participation

the Black Environmental Network has been at the forefront in challenging the racial exclusivity of environmental issues.

BEN was formed in September 1988 during and after a conference,
'Ethnic Minorities and the Environment', jointly sponsored by Friends of
the Earth and the London Wildlife Trust, but initiated by black activists
(Agyeman et al. 1991). It was born as a result of a series of happenings:
BEN's founder was involved with the Ethnic Minorities Award Scheme
(EMAS) of the National Council for Voluntary Organisations (NCVO),
which was part of European Year of the Environment (1987); networking
among black environmentalists, including the photographer Ingrid Pollard;
and interest from within the environmental movement in how to involve
people of colour in environmental campaigning and action.

EMAS had a grants policy which was general and inclusive; this was in
contrast to most environmental grants schemes of the time, which were
specific and exclusive. These tended to prescribe the kinds of projects
which the funders wanted to see, thereby reflecting a white middle-class
cultural interpretation of 'environment'. EMAS invited applications for
funding which raised environmental awareness; contributed to environmental improvement; and linked arts and environment. Agyeman et al.
note that 'ethnic minorities were poised to take part [in the environmental
movement] ... and that the character of some of their projects was of
particular interest in crossing boundaries – combining environmental with
cultural, artistic and social issues' (1991: 6).

A major funding request from groups of people of colour was for
countryside trips, showing that there is an interest in access. The Union
of Turkish Women in Britain, an early recipient of funding, reported that

> we then went to Hastings, 150 people enjoyed a delicious kebab
> barbecue up on the cliffs. Some of us sang and joked, others played
> football. In the late afternoon we went down to the seaside. It had
> been a long time since most of us had seen the sea since leaving
> Turkey. Cooking on an open fire reminded us of the good old days
> when we used to go on picnics back home.
>
> (quoted in Agyeman et al. 1991: 10)

This interest in visiting the countryside encouraged the Countryside Commission and English Nature to fund the countryside trips. It is interesting
to note that the two major donors to BEN were government agencies.
Both had concerns, resulting from arguments articulated by BEN, about
their inability to attract people of colour to their areas of activity; that is,
countryside access and nature conservation. This sponsorship resulted in a
firm dialogue between EMAS, and then BEN, on the one hand; and the
Countryside Commission on the other. The commission's 1991 publication,
Visitors to the Countryside, incorporated many of BEN's suggestions,
especially those concerning the need for positive imagery, and for targets

for the employment of people of colour as rangers. It dropped the latter suggestion in 1993. When questioned, the Commission implied that the situation was actually more subtle than this. It had raised the issue with those bodies employing countryside staff, for example, local authorities and the National Trust; now the onus was on these organisations to take action. The Countryside Commission shifted its focus to encouraging participation among young people of colour, in the hope that early enjoyment would lead to a desire to work in the countryside. We would suggest that this simply diffuses the situation, and is perhaps a case of racism via the 'back door'.

BEN and EMAS merged fully in 1990, after much collaboration, having in 1989 appointed a part-time development worker, Judy Ling Wong, who later became Director. With a London base (at NCVO), BEN began to set up 'Regional Forums' in major urban areas, which consisted of local environmental organisations and representatives from ethnic minority communities. The aim was to get people talking: the environmental organisations would see a different interpretation of 'environment', the ethnic minority representatives would be able to tap into the expertise of the environmental groups. The Regional Forums were renamed 'Local Groups' in 1992/93.

BEN's project was, in the early days, essentially ideological and political; it worked 'to enable the missing contribution of black and ethnic minority communities to come forward in environmental work' (Agyeman et al. 1991: 1). As such, BEN was one of the first 'environmental' organisations to push the links between environmental, equity and social justice issues. Many had done so on a global scale (for example, Survival International), but few, if any, were doing so in the UK context.

However, many people, when asked, think of BEN only as promoting access to the countryside among people of colour. This is due, in large part, to two articles: Coster's (1991) 'Another country' in the *Guardian*, and Deedes' (1992) mischievously (and originally) titled 'Another country' in the *Daily Telegraph*. The content of, approach of and conclusions to these articles need not concern us here; they reflect the papers' well-known political affiliations. However, what is of interest is that they were both large articles (three pages of the *Guardian* Weekend and one full page of the *Daily Telegraph*'s weekend section) which were the result of ideas from a small organisation run by people of colour (BEN). That they achieved such a large number of column inches reflects both the sensitivity of the issue in its challenge to established concepts of national identity (especially in Deedes 1992) and the effectiveness of BEN in getting the issue onto the middle-class agenda.

This has to be tempered against the comments made earlier, namely that BEN was and is more than a countryside campaign for people of colour. Similarly, television coverage of BEN, as the only black environmental organisation, and the only organisation which was at that time articulating a broader environmental agenda, was restricted to issues relating to countryside access.

The political project of BEN, according to Taylor (1993), is one of consensus. She notes that 'the government has created an atmosphere wherein environmental groups concentrate on being drawn into the inner circle rather than being critical of the process', and that 'the Black Environment Network is positioning itself for that possibility sometime in the distant future' (1993: 280). Taylor suggests that 'the introduction of a minority perspective into certain issues is sometimes met with resistance, discomfort, denial and efforts to delegitimate the voice with that new perspective' (ibid.: 270). Perhaps not surprisingly, BEN has faced a number of problems in progressing its ideas; these include its relationship with 'race' bodies, such as the Commission for Racial Equality (CRE), and its relationship with the media. In terms of its relationship with the CRE, BEN was seen initially as an irrelevance. When researching a programme for Anglia TV, on recreational access to the countryside, BEN was told, 'there is no issue'. Yet later, the CRE bought out two reports on ethnicity and the countryside: *Keep Them in Birmingham* (Jay 1992) and *Not in Norfolk* (Derbyshire 1994). Admittedly, these reports focus on people of colour living in the countryside, but to the authors' knowledge, no approach was made to BEN in researching these reports, nor was any acknowledgement of BEN's work made in them, despite BEN's clear lead in raising the issue of ethnicity and the countryside (Agyeman 1989a).

Has BEN's project been assimilated by white individuals and organisations, as hinted at by Taylor (1993)? To a large extent it has. During 1995, it had a white Chair, and only one black management committee member (who has since resigned). The others on the committee represented funding bodies and organisations with which BEN was working. To this extent, while it aims to reach (i.e. fund) people, communities and organisations of colour, it can no longer claim to represent their expressed interests in that their voices are no longer heard on the management committee. BEN's ability to launch a contestatory politics of resistance to the mainstream white environmental movement has been co-opted by these very organisations.

CONCLUSIONS

The response that there are few people of colour living in the countryside and therefore there can be no racism or question of equal opportunities is clearly an inadequate one. The geography of ethnicity, whether in terms of people of colour or white people, is not purely an urban phenomenon, but has resonances across rural spaces too. There is a growing body of work in geography and the environmental sciences which is beginning to address the associations between racism and an exclusive white British national identity, embedded in the countryside. Here we are not suggesting a neglect of urban issues; indeed, the question of urban-based recreational access to these spaces is as important as issues affecting those living in the countryside. A number of organisations working at the urban–rural interface are beginning to tackle this exclusion. For example, the British Trust for Conservation Volunteers (BTCV), which undertakes practical environmental action, has recently moved from being a very middle-class, conservative organisation to pushing an equal-opportunities programme, making contacts with many locally based groups run by and for people of colour. This proactive stance is one that other organisations could learn from.[13]

A recurring theme of this chapter has been the relationship between the alien and the native, in terms of both people and plants. We have begun to suggest that ideas of native and alien peoples are in some way strengthened by associations with ecological (scientific) discourse. For example, various ecologists, in favour of 'native' species, have suggested practices such as 'rhodo-bashing', to rid the countryside of 'alien' plants. The frightening and outrageous connection with racist practices is immediately obvious. The ecological 'alien' is being anthropomorphised in the popular imagination.

These issues are all tied together by images in popular culture which represent the countryside as, among other things, a white space. These images are being contested, for example in the work of Ingrid Pollard; however, more conservative media, such as Radio 4's *The Archers*, are beginning to recognise the salience of ethnicity in the countryside.

Where ethnicity is recognised it is always in terms of an ethnic Other to whiteness. Jane Goldsmith, a mixed-race African/European woman and author of the report *Staring at Invisible Women* (National Alliance of Women's Organisations 1994), tells of how, as a child, growing up in a village where there was nothing to do, but everyone came out to watch her do it! Paradoxically, blackness, although highly visible, remains invisible in terms of rural service provision. The countryside is regarded as being free from issues of ethnicity owing to its perceived whiteness, a colour which remains invisible. Studies of ethnicity in the countryside should turn towards constructions of whiteness, and how it is related to questions of power and exclusion as well as exposing its whitewashing as simplistic and false.

NOTES

1 Huge debate surrounds the use of terms such as ethnicity and 'race'. Here we use ethnicity for its self-ascriptive nature. We recognise the various criticisms of ethnicity (see, for example, Maynard 1994), but if, as Stuart Hall (1992)

points out, we accept the social and historical construction of 'race', then the term referring to this is ethnicity.

2 We have elected to use the term 'people of colour' in this chapter. It is used to describe people of Asian, African and African–Caribbean descent, avoiding the criticisms and homogenising tendencies of the term 'black' (West 1993; Modood 1988, 1992). We use it as a political signifier under which people of colour may combat racism, while highlighting awareness of differences between very diverse peoples.

3 For an exception see Bonnett (1992, 1993).

4 For exceptions see Bonnett (1992, 1993), Kinsman (1993), Smith (1993), Daniels (1992, 1993) and Wrench et al. (1993).

5 The Black Environmental Network employs a very inclusive definition of 'black' in its name, a definition which was itself heavily debated (Taylor 1993). Ling Wong (1993: 8) writes, the use of the term 'black' in our name is symbolic and is used to describe the common experience of all ethnic minority communities, including the less visible white minority communities such as the Polish, Greek Cypriots or the Irish'. The obvious conflict with our own rejection of the term 'black' is evident, and is further illustrative of the wider debate over terminology (Dyer 1993; Smitherman 1991).

6 In our discussions it was very easy to locate Rachel as white, but it proved very difficult to describe Julian. A 'person of colour' seemed too impersonal, while 'black' obscured other aspects of Julian's identity. This again points to the importance and difficulty of 'naming' for people designated Other, and the ease with which white people define themselves.

7 For example, see note 6 for our unease in defining Julian, but not Rachel.

8 Again, we are uneasy with ascribing labels which do not say all there is to say about a person.

9 Here, we concentrate on constructions of 'race'. We do recognise that other differences are inscribed into the rural–urban dichotomy, for example in terms of gender (see Davidoff, et al. 1976).

10 An exception to this is BEN, for which the environment was identified as a cause célèbre.

11 As discussed earlier, this connection between national culture and rurality may have racist connotations.

12 The use of sexual metaphor is perhaps an allusion to the racist stereotype of black male sexuality as a threat to white femininity (Ware 1992).

13 There are a number of other environmental organisations run by and for people of colour besides BEN; for example, the Black Environmental Action Group. However, for every positive example there appears to be a more negative one. For example, we contacted the Countryside Recreation Network, which co-ordinates information on countryside access, and the only research concerning ethnicity revealed by a quick search of its database was one regarding overseas visitors on a North Sea Ferry Passenger Survey!

REFERENCES

Agyeman, J. (1989a) 'Black-people, white landscape'. *Town and Country Planning*, December: 58 (12) 336–8.

—— (1989b) A snail's pace, *New Statesman and Society* 2 (35): 30–1.

—— (1991) The multicultural city ecosystem, *Streetwise* 7: 21–4.

—— (1993) 'Alien species', *Museum Journal*, December: 22–3.

—— (1995) 'Environment, heritage and multiculturalism', *Interpretation: A Journal of Heritage and Environmental Interpretation*: 5–6.

Agyeman, J., Warburton, D. and Ling Wong, J. (1991) *The Black Environment Network Report: Working for Ethnic Minority Participation in the Environment*, London, Black Environment Network.

Anderson, K. (1988) 'Cultural hegemony and the race definition process', *Society and Space Environment and Planning D* 6: 127–49.

—— (1991) *Vancouver's Chinatown: Racial Discourse in Canada 1878–1980*, Montreal 1 Buffalo, McGill Queen's University Press.

Banton, M. (1977) *The Idea of Race*, London, Tavistock.

Barker, G. (1994) 'Which wildlife? What People?', *Urban Nature* 2, 14–17.

Bell, D. and Valentine, G. (1995) 'Queer country: rural lesbian and gay lives', *Journal of Rural Studies*, 11 (2): 113–22.

Bolton College (1993) *The Bolton Initiative*, Bolton, Bolton College.

Bonnett, A. (1992) 'Anti-racism in "White" areas: the example of Tyneside', *Antipode* 24 (1): 1–15.

—— (1993) 'Contours of crisis: anti-racism and reflexivity', in Jackson, P. and Penrose, J. (eds) *Constructions of 'Race', Place and Nation*, London, UCL Press.

Bunce, M. (1994) *The Countryside Ideal: Anglo-American Images of Landscape*, London, Routledge.

Chambers, I. (1990) *Border Dialogues: Journeys in Postmodernity*, London, Routledge.

—— (1993) 'Narratives of Nationalism: being "British" '; in Carter, E., Donald, J. and Squires, J. (eds) *Space and Place: Theories of Identity and Location*, London, Lawrence and Wishart.

Cloke, P. (1994) '(En)culturing political economy: a life in the day of a "rural geographer" ', in Cloke, P., Doel, M., Matless, D., Phillips, M. and Thrift, N. (eds) *Writing the Rural: Five Rural Geographies*, London, Paul Chapman.

Cloke, P. and Milbourne, P. (1992) 'Deprivation and lifestyles in rural Wales–II. Rurality and the cultural dimension', *Journal of Rural Studies* 8: 359–71.

Cloke, P., Phillips, M. and Thrift, N. (1995) 'The new middle classes and the social constructs of rural living', in Butler, T. and Savage, M. (eds) *New Theories of the Middle Class*, London, UCL Press.

Coates, I. (1993) 'A cuckoo in the nest: the National Front and green ideology', in Holder, J., Lane, P., Eden, S., Reeve, R., Collier, U. and Anderson, K. (eds) *Perspectives on the Environment: Interdisciplinary Research Network on the Environment and Society*, Aldershot, Avebury.

Cohen, A. (1982) 'Belonging: the experience of culture', in Cohen, A. (ed.) *Belonging: Identity and Social Organisation in British Rural Cultures*, Manchester, Manchester University Press.

Coster, G. (1991) 'Another country', *Guardian* Weekend section, June: 4–6.

Countryside Commission (1992) *Enjoying the countryside: Policies for People*, Cheltenham.

Dabydeen, D. (1992) *Disappearance*, London, Secker and Warburg.

Daniels, S. (1992) 'Place and the geographical imagination', *Geography* 7 (337), Part 4.

—— (1993) *Fields of Vision: Landscape Imagery and National Identity in England and the United States*, Cambridge, Polity Press.

Davidoff, L., L'Esperance, J. and Newby, H. (1976) 'Landscape with figures: home and community in English society', in Mitchell, J. and Oakley, A. (eds) *The Rights and Wrongs of Women*, Harmondsworth, Penguin.

Deedes, W. F. (1992) 'Another country', *Daily Telegraph* Weekend section, 1 April: 1.

Derbyshire, H. (1994) *Not in Norfolk: Tackling the Invisibility of Racism*, Norwich, Norwich and Norfolk Racial Equality Council.

Di Leonardo, M. (1994) 'White ethnicities: identity politics and baby bear's chair', *Social Text* 41: 165–91.

Dodd, P. (1986) 'Englishness and the national culture', in Colls, R. and Dodd, P. (eds) *Englishness, Politics and Culture 1880–1920*, London, Croom Helm.

Doughty, R. (1978) 'The English sparrow in the American landscape: a paradox in nineteenth century wildlife conservation', *Research Papers 19*: 1–36, University of Oxford, School of Geography.

Duncan, J. and Ley, D. (1993) 'Introduction: representing the place of culture', in J. Duncan and D. Ley (eds) *Place, Culture, Representation*, London, Routledge.

Dyer, R. (1988) 'White', *Screen* 29 (4): 44–65.

—— (1993) *The Matter of Images: Essays on Representation*, London, Routledge.

Eaton, L. (1994) 'Turks delight in an Essex Sunday', *Independent London*, 19 August: 5.

Environ (1994a) 'The Asian community and the environment: towards a communications strategy', *Environ Research Report 16*, Leicester, Environ.

—— (1994b) 'The Afro-Caribbean community and the environment: towards a communications strategy', *Environ Research Report 19*, Leicester, Environ.

Fenton, J. (1986) 'Alien or native?', *Ecos* 7 (2): 22–30.

File, N. and Power, C. (1981) *Black Settlers in Britain, 1555–1958*, London, Heinemann Educational Books.

Frankenberg, R. (1993a) *White Women, Race Matters: The Social Construction of Whiteness*, London, Routledge.

—— (1993b) 'Growing up white: feminism, racism and the social geography of childhood', *Feminist Review* 45: 51–84.

Fryer, P. (1984) *Staying Power: The History of Black People in Britain*, London, Pluto Press.

Gilroy, P. (1993) *The Black Atlantic: Modernity and Double Consciousness*, London, Verso.

Giroux, H. A. (1993) 'Living dangerously: identity politics and the new cultural racism: towards a critical pedagogy', *Cultural Studies* 7 (1): 1–27.

Gundara, J. S and Duffield, I. (1992) Introduction, in Gundara, J. S. and Duffield, I. (eds) *Essays on the History of Blacks in Britain: From Roman Times to the Mid-twentieth Century*, Aldershot, Avebury.

Hall, C. (1992a) *White, Male and Middle Class: Explorations in Feminism and History*, Cambridge, Polity Press.

—— (1992b) Book review – *The Wages of Whiteness: Race and the Making of the American Working Class* by Roediger, D. R., *New Left Review* 193: 114–19.

Hall, S. (1990) 'Cultural Identity and Diaspora', in J. Rutherford (ed.) *Identity, Community, Culture, Difference*, Lawrence and Wishart, London.

—— (1992) 'New Ethnicities', in Donald, J. and Rattansi, A. (eds) *'Race', Culture and Difference*, London, Sage.

Haraway, D. (1991) *Simians, Cyborgs and Women: The Reinvention of Nature*, London, Free Association Books.

hooks, b. (1991) *Yearning: Race, Gender and Cultural Politics*, London, Turnaround.

—— (1992) 'Representing whiteness in the black imagination', in Grossberg, L., Nelson, C. and Treichler, P. (eds) *Cultural Studies*, New York, Routledge.

Howkins, A. (1986) 'The discovery of rural England', in Colls, R. and Dodd, P. (eds) *Englishness, Politics and Culture 1880–1920*, London, Croom Helm.

Jackson, P. (ed.) (1987) *Race and Racism: Essays in Social Geography*, London, Allen and Unwin.

Jackson, P. and Penrose, J. (1993) 'Identity and the politics of difference', in Jackson, P. and Penrose, J. (eds) *Constructions of 'Race', Place and Nation*, London, UCL Press.

Jay, E. (1992) *Keep Them in Birmingham*, London, Commission for Racial Equality.

John, E. (1988) *The Arkleton Trust Project*, St Albans, Youth Hostels Association.

Jones, K. (1992) 'Recreations', *Ten-8, Critical Decade* 2 (3), Spring: 96–105.

Kinsman, P. (1993) 'Landscapes of national non-identity: the landscape photography of Ingrid Pollard', *Working Paper 17*, University of Nottingham, Department of Geography.

—— (1994) 'Race, environment and nation: the Black Environmental Network', Paper presented to the 1994 Annual Conference of the Institute of British Geographers, Nottingham.

Lee, J. (1988) 'Care to join me in an upwardly mobile tango?', in Gamman, L. and Marshment, M. (eds) *The Female Gaze: Women as Viewers of Popular Culture*, London, Women's Press.

Liang, S. (1992) 'Images of the rural in popular culture, 1750–1990', in Short, B. (ed.) *The English Rural Community: Image and Analysis*, Cambridge, Cambridge University Press.

Ling Wong, J. (1993) 'Equally green', *School's Out*, Spring: 8–9.

Lowe, P. (1983) 'Values and institutions in British nature conservation', in Warren, A. and Goldsmith, F. B. (eds) *Conservation in Perspective*, Chichester, Wiley.

Lowe, P., Murdoch, J. and Cox, G. (unpublished) 'The civilised countryside: the place of the rural in British culture'.

Lowenthal, D. (1991) 'British national identity and the English landscape', *Rural History* 2 (2): 205–30.

McCrystal, C. (1995) 'A very filthy, very dirty and a very stinking place', *Observer*, 3 December: 11.

Malik, S. (1992) 'Colours of the countryside – a whiter shade of pale', *Ecos* 13 (4): 33–40.

Maynard, M. (1994) ' "Race", gender and the concept of difference in feminist thought', in Afshar, H. and Maynard M. (eds) *The Dynamics of 'Race' and Gender: Some Feminist Interventions*, London, Taylor and Francis.

Mingay, G. E. (ed.) (1989a) *The Rural Idyll*, London, Routledge.

—— (ed.) (1989b) *The Unquiet Countryside*, London, Routledge.

—— (ed.) (1989c) *The Vanishing Countryman*, London, Routledge.

Modood, T. (1988) ' "Black", racial equality and Asian identity', *New Community* 14 (3): 397–404.

—— (1992) *Not Easy Being British: Colour, Culture and Citizenship*, Stoke-on-Trent, Runnymede Trust and Trentham Books.

Morrison, T. (1992) *Playing in the Dark: Whiteness and the Literary Imagination*, London, Picador.

Mosse, G. (1985) *Nationalism and Sexuality: Respectability and Abnormal Sexuality in Modern Europe*, New York, Howard Fertig.

Mouffe, C. (1994) 'For a politics of nomadic identity', in Robertson, G., Mash, M., Tickner, L., Bird, J., Curtis, B. and Putnam, T. (eds) *Travellers' Tales: Narratives of Home and Displacement*, London, Routledge.

Murdoch, J. and Day, G. (1995) 'What is a rural community?', paper presented to the Migration Issues in Rural Areas Conference, Swansea, 27–29 March.

National Alliance of Women's Organisations (1994) *Staring at Invisible Women: Black and Minority Ethnic Women in Rural Areas*, London, NAWO.

Nicholson, B. (1987/8) 'Native versus alien', *London Wildlife Trust Magazine*, Winter (unpaginated).

Pollard, I. (1989) 'Pastoral interludes', *Third Text: Third World Perspectives on Contemporary Art and Culture* 7: 41–6.

—— (1993) 'Another view', *Feminist Review* 45: 46–50.

Puar, J. (1993) ' "It's great to have someone to talk to" – or is it?: The feminist process of research', paper presented to the Women and Gegoraphy Study Group, University of Central London, 16 November.

Roberts, L. (1992) 'A rough guide to rurality: social issues and rural community development', *Talking Point* no. 137, Newcastle upon Tyne, Association of Community Workers.

Savage, M., Barlow, J., Dickens, P. and Fielding, T. (1992) *Property, Bureaucracy and Culture*, London, Routledge.

Schoon, N. (1992) 'The barbarians in Britain's back yards', *Independent on Sunday*, 17 May.

Short, J. R. (1991) *Imagined Country: Society, Culture and Environment*, London, Routledge.

Sibley, D. (1995) *Geographies of Exclusion: Society and Difference in the West*, London, Routledge.

Smith, S. J. (1989) *The Politics of 'Race' and Residence: Citizenship, Segregation and White Supremacy in Britain*, Cambridge, Polity.

—— (1993) 'Bounding the borders: claiming space and making place in rural Scotland', *Transactions of the Institute of British Geographers* NS 18: 291–308.

Smitherman, G. (1991) ' "What is Africa to me?" Language, ideology and African American', *American Speech* 66 (2): 115–32.

Stanley, L. and Wise, S. (1993) *Breaking Out Again: Feminist Ontology and Epistemology*, 2nd edition, London, Routledge.

Taylor, D. (1993) 'Minority environmental activism in Britain: from Brixton to the Lake District', *Qualitative Sociology* 16 (3): 263–95.

Taylor, P. (1991) 'The English and their Englishness: a curiously mysterious, elusive and little understood people', *Scottish Geographical Magazine* 7 (3): 146–61.

Visram, R. (1987) *Indians in Britain*, London, Batsford.

Ware, V. (1992) *Beyond the Pale: White Women, Racism and History*, London, Verso.

West, C. (1993) 'The new cultural politics of difference', in During, S. (ed.) *The Cultural Studies Reader*, London, Routledge.

Williams, R. (1983) *The County and the City*, London, Chatto and Windus.

Wong, L. Mun (1994) 'Di(s)-secting and dis(s)-closing "Whiteness": two tales about psychology', *Feminism and Psychology* 4 (1): 133–53.

Wrench, J., Brar, H. and Martin, P. (1993) *Invisible Minorities: Racism in New Towns and New Contexts*, Monographs in Ethnic Relations no. 6, Coventry, Centre for Research in Ethnic Relations.

Wright, P. (1992) 'The disenchanted forest', *Guardian* Weekend section, 7 November.

12

ENDANGERING THE SACRED
Nomads, youth cultures and the English countryside

David Sibley

Michel Foucault observed that although [in the developed world?] there had been 'a certain theoretical desanctification of space . . . we may not yet have reached the point of a practical desanctification of space' (Foucault 1986: 23). His comment implies that, while in modern societies fewer spaces are endowed with sacred qualities than they were, say, in medieval Catholic Europe or in ancient Israel, material objects and spaces may still be represented as if they had sacred qualities. Whether a capitalist icon like the Coca-Cola machine which was violently assaulted in the film *Dr Strangelove* or spaces of power like Red Square in Moscow, there are things and spaces which are revered (by some) and, in Durkheim's terms, their special character 'is manifest in the fact that [they are] surrounded by ritual prescriptions and prohibitions which enforce [a] radical separation from the profane' (Giddens 1971: 107). As Law and Whittaker (1988: 173) recognised, in the representation of elements of the material world as if they were sacred, transcendent, core values are being celebrated. Thus, permanence and stability must be emphasised and, in the case of social space, it is necessary to suppress heterogeneity and to silence the discrepant – as Law and Whittaker put it, 'to prevent them from speaking in other ways'. Thus, there is a necessary distortion of geography and history. To mark off sacred spaces, to define boundaries which can be defended, they have to be imbued with special qualities.

In this chapter, I will suggest that current conflicts in and around the English countryside, concerned primarily with who belongs and who does not belong, reflect the sacred quality of the countryside as it has been represented by some rural communities and by the state. By looking at past instances of transgressions, I will try to demonstrate the consistency of arguments for the defence of rural space against diverse (allegedly urban) threats and then show how this singular view of the English countryside as a sacred social space is manifest in public order legislation introduced in 1994 by the Criminal Justice and Public Order Act. I will not want to leave the impression that there is a single, timeless boundary defining some quintessential English countryside but rather, that the existence of such a boundary is asserted during particular episodes when the interests of the more powerful groups in rural society appear to be threatened. In Kai Erikson's terms (Erikson 1966), there are periodic boundary crises when the sanctity and stability of an imagined 'rural community' appear to be endangered by the transgressions of discrepant minorities. At other times, alternative communities and various 'others' have been accommodated in dominant visions of the English countryside.

THE COUNTRYSIDE, NATIONALISM AND ABJECT OTHERS

Recently, there has been considerable critical discussion of the idea of *English* identity and, particularly, the ways in which constructions of Englishness are implicated in racist discourse (Gilroy 1990; Chambers 1993; Donald 1993). At many points in these discussions of national identity, the English countryside appears as a key signifier. Thus, in writing about the English landscape, David Lowenthal (1991: 213) suggests that the term conveys 'not simply scenery and *genres de vie* but quintessential national virtues'. While narratives of the English landscape may have found a place for towns, they have been primarily rural visions, portrayals of countrysides which, according to poets like Rupert Brooke and Steven Spender, men were dying for during the First World War.

This English countryside also had a central role in imperialist discourse.[1] The view of the English countryside as the source of essential Englishness has been sustained in the past by the manufacture of threats to its sanctity. Imperialism produced alien 'others' who might violate the boundaries of rural space, so, implicitly, they represented a danger to English values. For example, James Donald (1993: 175) argues that in Sax Rohmer's *The Mystery of Dr Fu Manchu* 'ideal rural peace' was made insecure by Fu Manchu, the sterotypical Oriental 'other' who brought home to the English countryside the threat posed by the racialised, colonised subject peoples of the British Empire. In the face of this imaginary threat, it was necessary to emphasise the homogeneity and social order of rural England. As Donald suggests, novels in the *Fu Manchu* genre 'take as [their] topic the abject and the schizophrenic: duality, transformations, the breaking of boundaries, the fragmentation of experience and identity. But their social function is aggressively centripetal. *Conjuring up chaos, they drive towards unity, order and wholeness*' (my italics).

These colonised peoples, as well as indigenous others such as nomads and working-class youth, are *abject*[2] in the sense that they instil feelings of nervousness and anxiety about the territorial security of dominant groups, but the fear of these abject others can never be removed entirely. The abject is constituted by an array of socially constructed 'others' which define the boundary separating the self and the social world, but they also define the boundary that separates a social collectivity from the larger society. There is a strong desire to expel the abject, but it hovers on the boundary of the self or the community, threatening but, at the same time, confirming identity. That which is abject can assume many forms. In relation to the body, the sense of abjection relates in the first place to residues – to sweat, scurf, faeces, and so on, which are socially defined as defiling. Bodily residues elide with abject 'others', however, and the same excremental language may be used to describe bodily defilement and abject social categories: dirty Gypsies, untidy, disorderly working-class families depositing their filth in the countryside, a common theme in the 1930s (Matless 1995), and so on.

Dominant rural communities in England have at different times identified threatening, abject others in the form of colonial and ex-colonial peoples, nomads, the urban working class and youth cultures. They have all defiled rural space, they all come from somewhere else. The pure, homogeneous countryside is opposed to the heterogeneous and disordered city in the same way that, in colonial discourse, the pure, white and well-ordered colonising nation has been opposed to the disordered, unruly societies in dependent territories (Dyer 1993). An awareness of urban and colonial 'others' cannot be removed, so, in this sense, they are always hovering on the boundary, but crises are precipitated when they transgress – for example, when they move into a purified rural space.

In making sense of rural transgressions, the actual geographies of migration are irrelevant. It is the belief of those who are fearful that abject 'others' originate somewhere else, in the city or former colonial territories, that is important. The danger has to be represented as an external one. This was clearly demonstrated in the House of Commons debate on the Criminal Justice and Public Order Bill in January 1994,[3] when one Conservative member, Sir Cranley Onslow, commented on hunt saboteurs in these terms:

> It is a disgrace that organized violence, deliberate provocation and physical assaults on people and animals have become an accepted way of life for a militant section of urban society. The saboteur movement has its roots not in the countryside but in the towns. Anyone who has seen busloads of Millwall [a London football team] supporters brought in to disrupt a hunt knows exactly what I am talking about.

Working-class plotland dwellers in the 1920s, newly mobile motorists and motor cyclists in the 1930s, New Age Travellers, ravers and hunt saboteurs in the 1990s have all been represented by antagonists as deviant groups of urban origin. As I will suggest later in this chapter, Gypsies are a special case, a minority which has been seen both to belong and not to belong in the countryside.

There have been a number of 'crises' in the twentieth century associated with unwanted groups, minorities who are discrepant according to elitist representations of rural England. The current crisis, which has been effectively created by the Criminal Justice and Public Order Act of 1994, is merely the most recent instance of a move by the state to confirm the boundaries of an imaginary rural community through the identification of a number of pariahs.

PUBLIC ORDER LEGISLATION IN THE 1990s

The Criminal Justice and Public Order Bill was published in December 1993. Although concern for the welfare of rural communities is not explicit in the wording of the legislation, it is apparent that it was transgressive minorities of supposedly urban origin who were the main targets. Rural peace and quiet and the legitimate activities of country-dwellers were threatened to such an extent that new legislation was required. The Home Secretary, Michael Howard, argued that the legislation would 'provide the country with the most effective system of criminal justice which it is possible for a government to provide', and he went on to explain that

> Part V of the Bill [dealing with public order] contains important measures designed to tackle the destruction and distress caused mainly to rural communities by trespassers. Local communities should not have to put up with, or even fear the prospect of, mass invasions by those who selfishly gather, regardless of the rights of others.

This comment was directed particularly at ravers, but he also maintained that those who opposed country sports (for example) had no right to 'trespass, threaten or intimidate'. In other words, they should confine their activities to campaigns to change the law rather than engaging in direct action. Although Howard did not mention them when introducing the bill, one of the main targets of the public order legislation was New Age Travellers, who had earlier been wished out of existence by both Margaret Thatcher and John Major.

While these were the specific irritants, the legislation had more general significance. It clearly reflects New Right ideology, which has two apparently contradictory strands: neo-liberalism and neo-conservatism. Neo-liberalism emphasises the virtues of individualism and freedom of choice in a market economy whereas neo-conservative thinking stresses strong government, hierarchies, rigid boundaries and discipline – the apparatus of control rather than individual freedom (Fyfe 1995). However, the Conservative government has employed these ideas selectively in order to serve class interests. Thus, the market is encouraged by de-regulation but, at the same time, both in regard to capital–labour relations and in the social sphere, those groups who threaten the unconstrained working of the market are subject to increasing regulation and discipline. This can be seen in legislation which has limited the powers of trade unions as well as public order legislation which restricts public protest, as in the case of animal rights protesters, for example. What Andrew Gamble (1988) described as 'the strong state', a state concerned with discipline and control, secures a space for the market to operate by regulating labour and criminal-ising dissident or troublesome minorities.

One element of the neo-conservative programme is regulation of the use of public space. Instituting strong rules about legitimate behaviour in public spaces clearly limits the freedom of groups opposed to government policies or capitalist practices. Protest often involves contesting space, as in protests over the export of live animals, which have brought animal rights activists into conflict with the police in port areas, or opposition to road construc-tion in both rural and urban areas. Prohibiting various activities in the countryside and criminalising some groups who wish to live in rural areas both result in the strengthening of boundaries. The rural does not neces-sarily become socially more homogeneous, but greater value is attached to homogeneity. Threats to its supposed cultural homogeneity register as transgressions, many of which are criminalised by the new public order legislation.

THE BILL

The public order clauses of the Criminal Justice and Public Order Bill dealt with two aspects of the social and cultural space of the English countryside, although largely in coded terms. First, it identified 'those who do not belong' and proposed means for their removal. Secondly, it identi-fied movements and migrations which needed to be curtailed. Some of the discrepant or not-belonging groups are also recognised as 'undesirable' migrants, but others, like many environmental protesters, are criminalised only because they move into the countryside, not because they are social outcasts in any other sense.

Although John Major's tirade against New Age Travellers at the Tory Party conference in 1992 may have contributed to the creation of new public order offences, this group was not mentioned in the bill. Rather, it cast a wide legal net which effectively criminalised all nomadic and semi-nomadic minorities. This is the combined effect of a number of measures. First, under Clause 61, Part II (Section 8(1)) of the 1968 Caravan Sites Act is repealed. This Act made it the duty of local authorities in England and Wales which had Gypsies and other Travellers 'residing in or resorting to' their areas to provide sites. Since, according to the Department of the

Environment's own census, only about half of Travellers 'on the road' had been accommodated on sites by 1992, the repeal means that a large minority with no legal stopping place will have no hope of changing their insecure, illegal status. The problem is made worse by the bill's proposal to repeal Section 70 of the Local Government, Planning and Land Act 1980, which empowered central government to pay the capital costs of site development. In an attempt to justify this repeal, the government has used the neo-liberal argument that Travellers could apply for planning permission for sites and finance them themselves.

Having removed the state's obligation to provide further sites for Travellers, the bill then proposed to strengthen the laws of trespass. In a clause clearly directed at Travellers (Section 45(1)), 'if the senior police officer present at the scene [believed] that two or more persons are residing on land and are present there with the common purpose of residing there for any period *and* ... that those persons have between them six or more vehicles on the land', the officer can direct them to leave. Failure to do so, or returning within three months, renders them liable to a fine or three months' imprisonment. In addition to this, under Section 46, if trespassers fail to leave the land and have been arrested for an offence, including an offence under Section 45, 'a constable may, after a reasonable time, seize and remove any vehicle on the land appearing to the constable to belong to or to be in the possession or under the control of that person (or persons)'.

Clearly, in combination, these clauses put any Travellers not on an official site in a very difficult position. They can either move and make themselves liable to another trespass charge in another location, or they can stay put, with the possible consequence of being imprisoned or fined and made homeless (since 'vehicles' in Section 46 includes caravans).[4] English and Welsh Gypsies, Irish and Scottish Travellers on the road in England and Wales, as well as New Age Travellers who were apparently the main target of the legislation, are put in a vulnerable position by this legislation. In a critical comment on the new trespass laws, Simon Fairlie (1994) observed that 'everyone has to be somewhere and it is in the interests of everyone that "somewhere" is not prison'. The virtue of the trespass laws, which had been on the statutes since the Middle Ages, was, according to Fairlie, that they treated trespass as a civil offence. This delayed an eviction while applications for a court order were heard, so there was at least some respite. Under the new legislation, nomadism is effectively outlawed.[5]

Apart from Travellers, the new public order legislation criminalises a disparate group of protesters, people who are most likely to be taking direct action in rural areas. As the Home Secretary made clear, this was clearly the intention of the government. Section 52 (1) is concerned with what are termed 'disruptive trespassers'; that is, people who 'trespass on

223

land in the open air and, in relation to any lawful activity which people are engaging in or are about to engage in on that or adjoining land in the open air [do] there anything which is intended by [them] to have the effect a) of intimidating those persons or any of them so as to deter them or any of them from engaging in that activity'. Under 52 (4), 'A constable in uniform who reasonably suspects that a person is committing an offence under this section may arrest him without a warrant.'[6] Then, under Section 53 (1), 'if the senior police officer present reasonably believes that a) a person is committing, or intends to commit, the offence of aggravated trespass on land in the open air or b) that two or more persons are trespassing on land in the open air and are present there with the common purpose of intimidating persons to deter them from engaging in a lawful activity or obstructing or disrupting a lawful activity', he may order them to leave. Failure to comply or returning to the land as a trespasser within seven days is an offence punishable by a fine or three months' imprisonment.

This section of the bill appears to be directed primarily at hunt saboteurs. However, the legislation also criminalises environmental protesters campaigning, for example, against road building or other intrusive development in rural areas. It may have been the perception of the authors of the bill that people who engaged in protests were primarily young city-dwellers, opposed to government policies on a wide range of issues, but, more particularly, as Michael Howard implied in his introductory speech in the House of Commons, people who would challenge the 'core values' of the rural community. Recent protests over live animal exports and road construction demonstrate, however, a much wider coalition than this, including people who might be considered a part of the government's own rural constituency. The wide sweep of the legislation is indicated by the experience of one of the anti-live export campaigners,[7] who notes that 'just standing peacefully at the roadside can now lead to protesters being remanded in prison' and that the placing of conditional police bail on arrested protestors prevents them from taking any further part in protests. What, during the Twyford Down appeal, Lord Justice Hoffman referred to as the 'honourable tradition' of peaceful protest in Britain is thus seriously eroded by an Act that was designed to curtail the activities of a few non-conforming minorities.

MOVEMENT CONTROL

The English and Welsh countryside is secured for those communities which the Home Secretary saw as having a legitimate place in rural society partly by controlling the movements of threatening minorities. Having identified discrepant others, legal powers have to be available to make possible their exclusion or expulsion from rural space.

Attempts to control movement as a means of policing space have a long history. The social composition of the medieval European city, for example, was regulated by expulsions and by imprisoning undesirables in the gates of the city, the gates acting as a 'purifying filter', as Markus (1993: 118) put it. Gypsies in Europe, from the Middle Ages until recently, have been prohibited from travelling, they have been removed from national space by transportation to colonial territories, or their movements have been restricted. Thus, in France the compulsory identification card for Gypsies, the *carnet anthropométrique*, was used by the police until the 1960s as a means of controlling their migrations. On a wider scale, the *propitska* or residence permit was used in the former Soviet Union to restrict migration to cities and the pass laws in apartheid South Africa were instituted to enforce the segregation of the black majority population.

In England and Wales, a recent attempt by the state to control movement was the restriction on picketing during the 1984–85 miners' strike. A government ban on secondary picketing required the police to stop miners travelling to other pits to support their striking colleagues, a practice which was subsequently used in the policing of New Age Travellers who were suspected of migrating to sites with contested symbolic meanings, such as Stonehenge. The Criminal Justice Act extends the powers of movement control and puts a considerable onus on the police to decide who should be turned back in the case of groups or individuals suspected of travelling towards the site of an event which the police believe is going to take place and which could be illegal.

In the section of the bill concerned with movement and migration, only one group of people, ravers, are identified. Under 'Powers in relation to raves' (Sections 47, 48 and 49), a number of measures are introduced which demonstrate the state's capacity to manufacture moral panics and boundary crises. Raves will certainly go out of fashion in the short term but, as Stanley Cohen recognised in his classic study of moral panics (Cohen 1972), society always finds it necessary to demonise some manifestation of youth culture. Raves may appear particularly threatening because they involve relatively large-scale movements of primarily urban youth to sites which are often in rural areas. Under Section 47 of the Act, a police officer of at least the rank of superintendent has the power to direct people to leave the land 'if he reasonably believes that ten or more persons, present on the land in the open air, are waiting [for a rave] to begin there' (47(2)b). Beyond this, if a constable in uniform believes that a person is on his way to a gathering for which an order under Section 47(2) is in force, he may a) stop that person, and b) direct him not to proceed in the direction of the gathering. This mode of control can be exercised within a five-mile radius of the site where the police believe that a rave might take place.

The same power is granted under Section 55 through an insertion in the Public Order Act 1986 in relation to 'trespassory assemblies'. This term appears to cover festivals which 'a) are likely to be held without the permission of the occupier of the land or [to be conducted] in such a way as to exceed the limits of any permission . . . or which may result b) in i) serious disruption to the life of the community, or ii) where the land, or a building, or a monument on it, is of historical, architectural, archeological or scientific importance, in significant damage to the land, building or monument'. Again, these are possible consequences of assembly which the police have to anticipate. A senior police officer has first to decide on the likelihood of an event being disruptive or damaging and then prevent the movement of people to the site if s/he reasonably believes that they are on the way to such an event.

These are some of the clearest demonstrations in the Act of the strong state in action, a state which uses the police to control space in anticipation of an illegal act. In one sense, the movement control measures hark back to the discredited 'sus' laws, which were used in particular against black people and Gypsies.[8] They certainly criminalise some people who were formerly accepted in some localities as a part of the community or as participants in traditional rural activities.

A MORE TOLERANT PAST?

It may be nostalgia, but there is some evidence that, prior to the introduction of the Conservative government's controls over public space, there was a greater tolerance of difference in some rural areas. For some minorities, particularly Afro-Caribbeans and people from the Indian subcontinent, this was clearly not the case.[9] Gypsies also have been discriminated against fairly consistently in rural areas for several centuries.[10] However, some groups who are targeted in the Criminal Justice and Public Order Act, notably New Age Travellers in a slightly different guise, were accepted or, at least, tolerated in some localities during the 1960s and 1970s. This is evident from contemporary accounts of fairs and festivals.[11]

There is a continuing tradition of fairs in some parts of England and Wales which provided a focus for countercultural movements during this period. In East Anglia, horse fairs, like those at Bungay and Barsham, which had traditionally involved Gypsies and other horse dealers, began to attract people with an interest in alternative technologies, crafts, mysticism, and so on. Some of the participants explicitly rejected mainstream values. In some cases, they had adopted a semi-nomadic lifestyle, living in tipis and converted buses. Nomadism then contributed to the emergence of a festival circuit which included Glastonbury, the Green Moon festival at Nenthead in Cumbria, and other sites in rural England. These gatherings complemented more specialised events, like the Comtek festivals, which were concerned exclusively with alternative technologies. Events were advertised in shops catering to the counterculture, particularly bookshops and wholefood stores, and they were supported by magazines which encouraged networking as well as providing intellectual underpinnings for the movement. *Undercurrents*, for example, publicised events as well as providing space for articles on anarchist philosophy. Contemporary accounts suggest that the festivals caused little friction with the local community, and some, like the East Anglian fairs, involved local people.

It may be the case that rising unemployment during the 1980s and growing disenchantment with the materialism of British society during the Thatcher years attracted more people to a travelling life. Gatherings got bigger and New Age Travellers emerged as a new, threatening minority. One crucial event which signalled a change in attitude was the 'Battle of the Beanfield' in 1985. As one Traveller recollected,

> The Beanfield was a big turning point for everyone, I think. They'd placed an injunction on a few named people, banning them from going within five miles of Stonehenge, but really I think it was set up from way back; a deliberate plan to give everyone a good hiding and decommission [the Peace Convoy].
>
> (Lowe and Shaw 1993: 70)

This attempt to restrict the movement of Travellers resulted in violent conflict and considerable publicity which was generally anti-Traveller, despite the injuries and damage caused by the police. Although there had been some large-scale police operations at gatherings before, notably at the People's Free Festival in Windsor Great Park in 1974 and at Nostell Priory, Yorkshire, in 1984, the police action at the Beanfield was the clearest indication that the state considered nomadism and the lifestyle of New Age Travellers to be inherently deviant. Criminalising them with the 1994 public order legislation was consistent with the government's pronouncements and with its support for strong policing during the previous decade.

DECONSTRUCTING THE STEREOTYPES

It is interesting that the threat to rural harmony, according to the state, comes from transgressive urban subcultures of fairly recent origin. Gypsies, the largest semi-nomadic minority in England and Wales, were not a major concern in the parliamentary debates on the Criminal Justice Bill even though they would be seriously affected by the new laws on trespass and the repeal of Part 2 of the 1968 Caravan Sites Act. This may be because Gypsies have a long-established place in the English countryside. Of exotic origin and distinctive culture, romanticised Gypsies 'belong' in the rural landscape, along with thatched cottages and fox hunting. A closeness to nature has given wagon-dwelling Gypsies membership of the rural community, particularly as there are now very few wagon-dwellers left. It is urban Gypsies, living deviantly in chrome-trimmed trailers on derelict land, who have become the main target of abuse. Urban Gypsies are not 'real Gypsies'; there is no place for them in the city, according to popular constructions of people and place. In fact, hostility towards Gypsies shows no particular spatial pattern. It has been just as strong in rural, suburban and inner-city areas, but they have been protected to some extent by the romantic myth.[12]

New Age Travellers, by contrast, conjure up no romantic associations. They cannot claim exotic origins and they are not widely associated with the benign countercultural movements of the 1960s and 1970s, even though the world-view of some Travellers, like the Dongas Tribe, is rooted in the anarchism and ecological thinking of this period.[13] Urban New Age Travellers are castigated by Conservative and Labour politicians for aggressive begging, and an image of dependency combines with popular associations with the drugs scene, migrancy and dirt in a general sense to create a pariah group. Yet, as Lowe and Shaw's ethnographic studies demonstrate, the origins of New Age Travellers, their attitudes to drugs and alcohol, to welfare, and to the countryside, are quite diverse. The ascription covers a loose grouping of people who, for various reasons, have decided to travel or, if sedentary, to live without many of the material comforts of mainstream society. Unfortunately, ethnographies do little to dispel established negative images, particularly when the latter are reinforced by central government as a part of a larger political programme. New Age Travellers provide an ideal target for a government which wishes to foster the myth of a hard-edged, culturally homogeneous countryside in order to maintain the electoral support of its rural constituencies. Threats from the margins have to be cultivated.

CONCLUSION

The social composition of the English countryside is more complex than is suggested by those who wish to protect 'the rural community' from dangerous outsiders. Just as idealised and romanticised representations of English rural landscapes have no place for chicken factories, gravel pits, electricity pylons and council houses, so the representations of rural society articulated by its protectors have no room for Travellers, factory workers or ethnic minorities. These groups remain invisible in these versions of rural life.

In a densely populated, highly industrialised society where many rural residents have a lifestyle which is barely distinguishable from that of city-dwellers, 'the rural' as a social and spatial concept appears more distinctive if it is seen to be subject to threats from mobile urban minorities. Thus, at a time when the government sees that it can gain political advantage from defending rural interests which are, in fact, the interests of its traditional supporters living in rural areas, it serves its purpose to amplify the threats represented by various discrepant others. Those who inject an element of cultural difference into rural communities or who question the ethics of traditional practices in the countryside, such as fox-hunting, are criminalised. Putting up the barriers and manufacturing a boundary crisis is a strategic move which is a recurrent feature of national politics in situations where a government is concerned about internal dissent. This was evident in the 'Red scares' in the United States in the 1920s and again during the 1950s at the height of the Cold War, and in Britain following the introduction of the 1905 Aliens Act, when xenophobic politicians cultivated a fear of immigrants. The public order clauses of the Criminal Justice Act are concerned with the policing of interior spaces rather than national borders but the intention is the same. The legislation is designed to stifle dissent and to homogenise space and society (although the effect may be quite the opposite).

The English countryside has an important symbolic role in this manoeuvre because of its nationalistic associations. If it stands for essential Englishness, then transgressions by urban minorities constitute a threat to the fabric of English society. Ravers, hunt saboteurs and New Age Travellers are aliens whose incursions have to be resisted. In the past, this kind of reaction to difference has been episodic, and it is possible that the state's urge to use the law will diminish if threats other than those posed by rural nomads and youth cultures become more pressing. In any case, the cost of policing rural space may prove excessive, the law may be successfully challenged in the courts, or the value of diversity and difference may be appreciated in a more humane future.

NOTES

1 The xenophobic nationalism of the political right is supported by a vision of an unsullied and picturesque English countryside in the popular media. For example, an issue of *This England*, which claims to be 'Britain's loveliest magazine', expressed the following sentiments about exile: 'My days out here I am spending / In crying for the moon / For the sights and sounds of England / And summertime in June / And the scent of English roses / On an English afternoon.' These poetic sentiments accompanied a picture of a thatched cottage in Hampshire.

2 My comments on *the abject* and *abjection* are based in Julia Kristeva's writing (Kristeva 1981) and a dictionary essay by Elizabeth Gross (Gross 1992: 198). For a fuller account of the socio-spatial significance of abjection, see Sibley (1995).

3 I refer to the bill and the Act interchangeably, because the public order clauses of the bill became law with no major amendments. The quotations from debates are all taken from *Hansard*, 11 January 1994.

4 In the first legal challenge to these sections of the Act, in a High Court judgment it was ruled that local authorities had an obligation to take account of the welfare of Travellers before evicting them. In quashing an eviction order against New Age Travellers at Crowborough, East Sussex, Mr Justice Sedley

commented that local authorities had a responsibility to take account of 'considerations of common humanity, none of which can properly be ignored when dealing with one of the most fundamental human needs, the need for shelter with at least a modicum of security' (*Guardian*, 1 September 1995). The Act takes no account of these needs.

5 One of the objectives of Part 2 of the 1968 Caravan Sites Act was to facilitate the continuation of a nomadic life, although there were never enough sites built to make this possible. For an account of varying attitudes to nomadism as expressed in parliamentary debates during this period, see Sibley (1981: 95–103). Some New Age Travellers have suggested that rural France provides more hospitable spaces for travelling than England and Wales (Lowe and Shaw 1993).

6 Only one gender was recognised in the Bill.

7 This information comes from a letter to the *Guardian*, 19 September 1995.

8 As an example of the use of the 'sus' law in the 1970s, members of one Yorkshire Gypsy family were kept in police custody overnight on suspicion of being about to scavenge on a refuse tip, contrary to the 1936 Public Health Act.

9 The black photographer Ingrid Pollard and Julian Agyeman of the Black Environmental Network have raised awareness of this issue. A few local authorities have registered some concern about the absence of black minorities from the countryside. For example, Greater Manchester Countryside Unit held a seminar at the Bolton West Indian Community Centre on 'Minority Communities in the Countryside', in 1991, and Sheffield City Council, together with the Countryside Commission and the Royal Town Planning Institute, organised a conference on ethnic minorities and the countryside in 1992.

10 Several cases of evictions of rural Gypsies in nineteenth-century England are documented by Mayall (1988).

11 There are some accounts of wholesome rural fairs and festivals in the 1970s in the magazine *Undercurrents* and in Barnes (1983).

12 The right of 'real' Gypsies to roam the countryside was asserted in debates in the British Parliament on the Moveable Dwellings Bill in 1908 and 1911. Similarly, in exchanges in Parliament in 1952, the Minister of Housing and Local Government implied that Gypsies belonged in the countryside (Sibley 1981: 93; 1995: 104–5).

13 As 'Sam' of the Dongas Tribe put it, 'what the Dongas Tribe is about is reclaiming the land and trying to re-establish a vital tradition, the tradition of nomadism and living on the land, living and working alongside nature as opposed to working against it, trying to conquer and destroy it' (Lowe and Shaw 1993: 113).

REFERENCES

Barnes, R. (1983) *The Sun in the East: Norfolk and Suffolk Fairs*, RB Photographic, Windetts Kirkstead, Norfolk.

Chambers, I. (1993) 'Narratives of nationalism: being British', in E. Carter, J. Donald and J. Squires (eds) *Space and Place: Theories of Identity and Location*, Lawrence and Wishart, London, 145–64.

Cohen, S. (1972) *Folk Devils and Moral Panics*, MacGibbon and Kee, London.

Donald, J. (1993) 'How English is it? Popular literature and national culture', in E. Carter, J. Donald and J. Squires (eds) *Space and Place: Theories of Identity and Location*, Lawrence and Wishart, London, 165–86.

Dyer, R. (1993) *The Matter of Images: Essays on Representation*, Routledge, London.

Erikson, K. (1966) *Wayward Puritans*, Wiley, New York.

Fairlie, S. (1994) 'On the march', *Guardian*, 21 January.

Foucault, M. (1986) 'Of other spaces', *Diacritics* 16 (1), 22–7.

Fyfe, N. (1995) 'Law and order policy and the spaces of citizenship in contemporary Britain', *Political Geography* 14 (2), 177–89.

Gamble, A. (1988) *The Free Economy and the Strong State*, Macmillan, London.

Giddens, A. (1971) *Capitalism and Modern Social Theory*, Cambridge University Press, Cambridge.

Gilroy, P. (1990) 'The end of anti-racism', *New Community* 17 (1), 71–83.

Gross, E. (1992) 'Julia Kristeva', in E. Wright (ed.) *Feminism and Psychoanalysis: A Critical Dictionary*, Blackwell, Oxford, 194–200.

Kristeva, J. (1981) *Powers of Horror*, Columbia University Press, New York.

Law, J. and Whittaker, J. (1988) 'On the art of representation: notes on the politics of visualisation', in G. Fyfe and J. Law (eds) *Picturing Power: Visual Depictions and Social Relations*, Sociological Review Monograph 35, Routledge, London.

Lowe, R. and Shaw, W. (1993) *Travellers: Voices of the New Age Nomads*, Fourth Estate, London.

Lowenthal, D. (1991) 'British national identity and the English landscape', *Rural History* 2 (2), 205–30.

Markus, T. (1993) *Buildings and Power: Freedom and Control in the Origin of Modern Building Types*, Routledge, London.

Matless, D. (1995) The art of right living: landscape and citizenship, 1918–39', in S. Pile and N. Thrift (eds) *Mapping the Subject: Geographies of Cultural Transformation*, Routledge, London, 93–122.

Mayall, D. (1988) *Gypsy-Travellers in Nineteenth Century Society*, Cambridge University Press, Cambridge.

Sibley, D. (1981) *Outsiders in Urban Societies*, Blackwell, Oxford.

—— (1995) *Geographies of Exclusion: Society and Difference in the West*, Routledge, London.

13

'I BOUGHT MY FIRST SAW WITH MY MATERNITY BENEFIT'

Craft production in west Wales and the home as the space of (re)production

Clare Fisher

INTRODUCTION

This chapter considers the imaginary and different rural spaces inhabited by craft production. This form of production is thoroughly woven into a series of text(ure)s of rural representations of the past, and for the first part of the chapter I draw on a selection of these. It is an emotional and possibly spiritual association, hinted at here by the words of Long and Whatmore, who suppose that 'The association between crafts, artisanal activities, self provisioning and family enterprise, and rural localities runs deep (Long 1984). Interpretations of one reinforce interpretations of the other' (Whatmore *et al.* 1991: 1). Through an exploration of some of these interpretations, I hope to illustrate ways in which associations of rurality and craft production converge, reinforce and compound each other. To do this, and also to introduce craft production in rural Wales, I turn to the work of Iorwerth Peate, who alongside others strove to realise that these different spaces and places of production subvert many expected norms. This subversion forms the heart of my chapter.

As Long writes, the marrying of craft production and rural localities pivots on the idea of 'self provisioning and family enterprise', portrayed as peculiar social relations of production based near or in the home. In the second half of this chapter, by drawing upon my research with craft makers living and making in west Wales, I hope to catch glimpses into such spaces of (re)production.[1] To interleave the first and second sections of the chapter I introduce briefly the term 'livelihood'. It is a term which conjures up, in part, the relations of that (re)productive space. The makers' words and works display some of the issues that they face on a day-to-day basis as they struggle to engage in craft-based production from within or near the home in rural west Wales. Theirs is a struggle which contests the conventional spatial and cultural dividing apart of production from reproduction, work from home, while reintroducing the notion of diverse and different forms of production taking place in a landscape commonly portrayed as dominated by agriculture in terms of its productivity (Cloke 1989: 177). With particular, but not exclusive, reference to the words, work and lives of women, I will also examine how craft makers aim both to make from and to re-create the spaces of (re)production, opening out the idea of 'family enterprise' and how it may operate on a day-to-day basis. Although I have often found that within these craft households the use of space is shaped as a reaction against what are for craft makers 'mainstream' routines such as 'going out to work', this space is also upset by somewhat ironic references back to those segregated mainstream productive and reproductive domains previously rejected. To invoke a sensation of this troubled and in-between space, where I feel craft makers live and work from, I borrow a line or two from a paper by Shurmer-Smith. Although Shurmer-Smith was referring to another author's experience, that of travelling on an ocean liner, it is her touching of this in-between or liminal space that is important here. She writes:

> The space between countries, the transitory community, the removal from the structures generated by work, home, extended kinship, politics – the ship as Utopia. The space of difference between unacceptable worlds. . . . What I wanted was something between the two, which does not exist either as a word or a recognised condition.
> (Shurmer-Smith 1992: 6, 1)[2]

This catches the idea of sailing 'the between' of space, between the 'same' and the 'other', 'production' and 'reproduction', between the stable continents forming the charted but 'unacceptable worlds' which craft makers have resisted. Yet at the same time they often have to look towards these continents to find some sort of order and stability. There is hence congruence with both worlds, yet neither is simply implied by the author of the chapter's title. This craft maker spent her maternity benefit on a saw, a dangerous tool of production, but in so doing she was in part fulfilling an underlying desire, as an artist and also as a single mother, to be at home with her children while they were young.

CRAFT PRODUCTION AND RURALITY – THE REALITY?

In this next section I draw on the 'rural imagination' of first Iorwerth Peate and then William Morris, whose work provides an introduction to their experience of the push to 'other' craft production. Both men, working and writing during the late nineteenth and early twentieth centuries respectively, at the cusp of tradition and modernity, fought against what was considered to be the crafts' 'natural' demise, railroaded by the 'belief in the brilliance of Fordism' (Weiss 1988: 7). Such progress, they believed, did little more than divest the countryside of its diversity and production of its humanity.

In order to contextualise the work of Peate, I draw upon the writing of Pyrs Gruffudd, who has eloquently uncovered the 'brilliance' of Peate's and a number of his contemporaries' rural imaginings:

> the rural imagination presented here is not an uncomplicated romanticism. It is, rather, a dynamic engagement with a 'place on the margin' to borrow from Rob Shields (1991). The rural becomes almost a liminal zone which is seen as occupying ground between tradition and modernity and the societies they represent. . . . [This] challenged dominant ideas of 'progress': ideas based on industrial capitalism and urban life.
>
> (Gruffudd 1994: 61, 73)

Iorwerth Peate was one of H. J. Fleure's[3] first postgraduates at the Aberystwyth School of Geography and Anthropology during the early part of the twentieth century. As the son of a village carpenter he identified closely with a self-sufficient rural community, one in which the craft maker was a central figure. Meeting the everyday material requirements of his (or her) neighbours, the craft maker was supposedly producing artefacts in harmony with both his society and local 'nature'. Working and writing from the 1920s onwards, Peate was experiencing not only the aftershock of industrialisation, a time epitomised by the demise of *'the great diversity* of a working rural society' (Williams 1990: 14), but also, and relatedly, nationalism (Peate 1932: 244, 245).[4] Peate's work and writings reflect his intrinsic belief that all forms of production, particularly the manufacture of goods, should be situated within the *Welsh* countryside. Both agricultural production and manufacture, he maintained, were integral to a vital countryside and community. Peate was fiercely opposed to the divisioning of agriculture from manufacture, 'country' from 'town', arguing that this effectively restricted the diversity of both. He believed that this was a reckless divisioning of space, just one further facet of a broad sweep of colonisation, the imposition of an English interpretation of the countryside onto a Welsh one. To him, the direction of such changes was alien to and destructive of Wales as a nation (see Peate 1943).

Peate's imagined rural society became one which provided a repository for 'difference, diversity and tradition' (the Welsh language included; see Peate 1933). The Museum of Welsh Folk Culture, at St Fagans near Cardiff, represented his attempt to re-create and to conserve[5] his particular interpretation of rural space. Yet Peate's work was (and is) easy to dismiss as no more than sentimental musings, museum pieces, symbols of the *undeveloped nation*, throwbacks to a rural past which in reality had little connection to a progressive Fordist nation.

The treatment that his work has received closely parallels that meted out to others accused of nostalgically fabricating a rural past, among them craft makers such as William Morris. For Morris, an imagined Arcadian rurality not only represented deliverance from the dehumanisation (such as the deskilling of craft workers) and despoliation of 'Nature' wrought by 'the machine age', but also was the 'utopian' basis of a new social movement and society. Morris pictured a revolution in which 'the town grew more like the country, yielded to the influence of their surroundings and became country people' (Morris 1891: 244). Gillian Naylor's comments on Morris and certain of his contemporaries seeks to expose their apparent naiveté, an inability to understand that

> their concern with the values of the past might so conflict with the realities of the present that their work would be rendered useless. . . . [N]one of them were in a position to grasp the nature of the forces that were revolutionising society. Theirs was a personal and a subjective approach, and although they came to appreciate intellectually the fact that the machine was the normal tool for our civilisation, they were unable, or unwilling, to absorb, as Lewis Mumford has put it, 'the lessons of objectivity, impersonality, neutrality, the lessons of the mechanical realm'.
>
> (Naylor 1971: 8)

RATIONALISING RURALITY

Unlike the case of urbanity, rurality and causality have never really mixed. The 'urban' domain has seemingly always held sway, such that, reversing Morris's view, the glittering town and belching factory appear to corrupt the innocent countryside (see, for example, James 1991; Pahl 1966).[6] Gathering alongside the selected examples sketched above, many others within community studies[7] and rural geography have presented imagined irrational ruralities, rather than rational real ruralities. The criticism of unnatural sentimentalism within such work culminated in Hoggart bluntly demanding 'LET'S DO AWAY WITH RURAL', in a paper written in 1990. He reasoned that 'the undifferentiated use of "rural" in a research context is detrimental to the advancement of social theory' (Hoggart 1990: 245). Although he kindly allowed 'lay' people to continue with their use of the term, writing, 'Let me be clear; I am not calling for the abandonment of the word "rural" in everyday expression' (ibid.), the outcome, as Whatmore states, is still ultimately to do away with vital clues:

> crucial dimensions of the legitimation, accommodation and contestation of the established power relations and hence, to the process of social change . . . [one] marginalised by theories of false consciousness, or irrationality, and in a vain faith of objectivity, and hence the

superiority of the scientific account.... [Hoggart's urging of rural researchers to forget about theorising in terms of 'rurality' ends up] silencing voices that have contradicted that account and stood in the way of progress.

(Whatmore 1990: 255)

However, as the work carried out by, for example, Crouch and Ward (1984), Halfacree (1993) and Philo (1992) has importantly demonstrated, there is a pressing need to reclaim such 'lay discourses' whose outplaying is to be understood as being as much a part of the workings of the world 'out there' as 'in here' (the workings of the academy) (see Katz 1992). The result is also to acknowledge the possibility that evacuated, theoretically driven senses of rurality are likewise influenced by more imaginative senses of space, whatever they may be, wherever they may spring from (Said 1978: 3; Cloke 1994: 153). Through my research I attempt to excavate some of these local, positioned senses of 'rurality' possessed by craft makers. In the course of my conversations with them I have found that their sense of, and desire to live in, rural west Wales is highly important to them. For those makers who have moved here, their choice of place was often driven by a sense of personal history and imagination.[8] As I go on to look at in greater detail in the second section of this chapter, by this rural space, which in part 'allowed'[9] them to engage in their different forms of production. Yet makers would not only dismiss their own imagined senses of rurality with comments like 'you'll think me mad if I mention this' but also would parody my interest in it, and would sometimes make comments along the lines of:

'We're not just selling a craft product, we are selling a whole way of life, all this. [She waves her arms towards the window and the expanse of the countryside beyond.] People come here and say, "Oh isn't this wonderful, I'd love to do what you're doing." But they don't mean it, they don't want to give up their comfortable house, nice car, nice job and come here ... we sort of do it for them.'
(Jan, toy-maker, speaking from the workshop connected to their home/smallholding)[10]

LIVELIHOODS, IMAGININGS AND MARGINALISATION

Before I move on to the second section, let me provide a further set of prefatory remarks. Here I wish briefly to introduce two further imagined ruralities loosely connected with craft production, yet closely centred on the issue of household-based production. The first is found in the writings of certain early feminists, Hayford (1974)[11] and de Beauvoir (1988) among them, who in their backward glances to a less 'commodified' rural existence spied somewhere, far off, a precapitalist society where women were not confined to a seemingly diminished reproductive role, but played a fully productive/economic one as well. This was a role that perhaps existed prior to the divisioning of space, 'of work' from 'home', as the distinctive and determining patterning of the capitalist mode of production. For certain feminists, as Bennett writes, this medieval, precapitalist mode of production had a distinct appeal, with women spinning and men weaving in the same spaces. In this 'golden age' women appeared to enjoy rough equality with men. Yet such idyllising is apparently deeply deceptive, as Bennett continues:

> Wives were expected to defer to their husbands in both private and public.... [Men] controlled both domestic and community affairs ... and [women] were considered to be fully dependent on their heads of household ... inheriting only in the absence of brothers and receiving lower wages for their work.
>
> (Bennett 1987: 6, 7)

A second imagining of importance here can be found in a more contemporary body of work, that of the so-called eco-feminists. They find strong inspiration in, and often make shameless recourse to, such a 'golden age' (which reflects aspects of Hayford's work; see Hayford 1974). Within this field, Merchant's work (1989) follows the themes of an organic (rural), female order being distorted by a mechanical, scientific, male order. Her portrayal of craft production, hence, holds to the belief that prior to the scientific revolution which, under capitalist patriarchy brought in rational mechanical order, hand tools remained within the 'grasp' and repair of both men and women, as well as being utilised within the frame of 'nature's' own constraints and contingencies.

Through home-based craft production, it is believed by de Beauvoir, eco-feminists and others, women have been able to achieve some greater sense of equality, engaging in production and doing so in an environment where making by hand had not yet been domesticated.[12] Such writers envisaged a sort of holistic utopia where production and reproduction are not spatially divided off. Their work, like Peate's, has now been somewhat discredited, marginalised for a persistent reference back to some utopian rural society, overshadowed by both rational academic and productive thinking which has concurred with both a spatial and mental dividing off of production and reproduction, 'work' and the 'home' (Whatmore 1991).

Conventionally there has been not only a counterpoising of the domains of production and reproduction but also a privileging of productive, economic work over the reproductive, work over home. Whatmore admits that academic and theoretical positioning of home-based production has generally conformed to this patterning and so has colluded in its marginalisation. In the context of academic terms and debates in regular currency,

Table 13.1 Conventional divisions associated with the domains of production and reproduction

Production	Home-based production livelihood	Reproduction
Large-scale industrial		Small-scale
Efficient		Inefficient
Rational		Irrational
Male		Female
Public		Private
Serious		Trivial
Developed		Undeveloped
Capitalism		Community
Culture		Nature
Global		Local
Urban		Rural

household-based production seems traditional or trivial, petty or transitional or domestic (Whatmore *et al.* 1991: 3). Through its apparent triviality or smallness next to capitalism proper, and through its associations with the 'home', domesticity and things (repetitively) reproductive, craft making has been viewed as marginal to the important working of contemporary society and craft makers themselves have often been socially and culturally marginalised in very real (and sometimes threatening) ways – themes which have been brought together and schematically illustrated in Table 13.1.

It has been through Whatmore's work, and that of other feminists, that the notion of livelihood has been recovered and its exploration begun (Whatmore 1991). Paradoxically, their 'imagined rurality' as a different space bearing different relations of production, achieves some sort of global resonance with those writing on and from so-called developing countries. This work has given greater credence to our understanding of different strategies of (re)production reweaving the domains of production and reproduction through the concept of *livelihood* (see, for example, Moore 1988; Shiva 1989). Whatmore writes:

> [This] means a recasting and interweaving of the processes and complex geographies of production and reproduction. Most importantly it recasts these processes as practical and meaningful, rather than abstract and logical ... in ways which challenge the boundaries of 'home' and 'work'.
>
> (Whatmore 1991: 147)

In the following section I will discuss the everyday practicalities of how certain makers compensate for, and indeed work with and sometimes subvert such marginalisations, 'recasting and interweaving of the processes and complex geographies of production and reproduction' (Whatmore

1991: 147). What I hope is illustrated is the 'complexities' of these reweavings, and, between the 'failings' and occasional 'triumphs', I hope that the makers' sense of attempting to engage with differences, in this case their chosen form of (re)production, is revealed to be potentially transformative, if but for an instant (Katz 1992: 498).

CONTEMPORARY HOME-BASED PRODUCTION IN WEST WALES

In the remainder of this chapter I explore makers in west Wales, reconstructing their particular and sometimes peculiar use of their home as a space of (re)production. I show how workshop space and work space within the home are negotiated, and how between the two, the home and workshop, there can be a considerable degree of confusion, evolution and reabsorption (of these spaces) as makers struggle to create space from which to make effectively. Of the thirty-one craft makers I interviewed, twenty-four were engaged in manufacturing their products either directly from home or from a workshop close by. The remaining seven owned or, more commonly, rented a workshop at a distance from their home. I begin with some of their thoughts illustrating why makers engage with craft production as conducted in this way.

Various statistics suggest that more women than men engage in craft production; the rounded ratio often quoted is something like 60 : 40 women to men (Knott 1994: 11) – although I would not go so far as writers such as Droste in making craft a 'woman's profession' (Droste 1990: no pagination). Undoubtedly though, women (and I would say some men as well) do move into craft production as a way of seriously avoiding both the structuring and the control of individual time and space that the nexus of capitalism and patriarchy imposes. These three makers, all women, I think capture the spirit of what Droste had in mind here:

'It's a way of women by-passing the structure, where they are excluded because they have children or they don't get higher education or things like that, and in fact women are incredibly good at business and they are very good at making things. . . . I think it's an interesting progression for women . . . it's certainly something because there is not a defined career structure. You can actually weave in and out, get what you want out of it and be successful, you know that's a really good development and certainly in crafts you can do it.'

(Jan, toy-maker)

Production from the home is a form of resistance and is about reclaiming control:

'I do it [i.e. craft production] because I can control my life. When I first moved down here I could have got a job as an occupational therapist, but that would be going out to work and doing what other people told me to do.... You know today I have been painting the house over these few days because it is bright and sunny.... My orders I do at night... I like to be able as much as possible to control my life, and craft is one way.'

(Louise, pyrographer)

'We've done things because we don't like the mainstream. People have not dropped out but said, "Look, I don't want to be involved in going in to a job nine to five, clocking in and clocking out, getting money and that's your life and if someone decides that's it, they no longer want you, you are given a week's notice, maybe a week's pay, if you're lucky and that's it. Termination!" ... you're not in control of what goes on. People want to be in control of their lives.'

(Sam, marquetry maker)

Being in control, as these quotes suggest, is not merely an issue of combining production with reproduction, there is also a sense of a repossession of space, and because of that a regaining or even 'recrafting' of identity:

'[with craft we are able] to express maybe a different sexuality, or work ethos or mentality which is acceptable in the crafts. [Women] can become furniture-makers, potters... sometimes so-called men's trades.'

(Rhian, silversmith)

Interestingly, on entering 'men's' or 'women's' trades or crafts, there can be a sort of cross-dressing against the stereotyped and often unacceptable conventions that particular occupations may impose. This can include reacting against very real discomforts with work situations such as wearing particular clothes and playing particular roles (as is the case for this furniture-maker, who used to teach):

Katrin   'I cringe now to think what I used to wear to work, it just wasn't me!'
Author   'So what did you used to wear?'
Katrin   'Oh, these skirts.... Oh God, I cannot bear to think about it!'

(Katrin, furniture maker)

THE HOME AS THE SPACE OF (RE)PRODUCTION

One weaver, whose basement was full of looms and stairs stacked high with bobbins, said that the only room which for the time being remained relatively 'sane' was the room we sat in for the interview. It was a sort of hallway cum sitting-room cum living space and the remnants and artefacts of production could be seen lapping menacingly at the doors. In this context, productive space seemed to have an overwhelming tendency to absorb the reproductive. And again, reproduction in the purely biological sense is not implied. Indeed, although this weaver is gay and he and his partner have no children, I would still consider the space of the factory–home (re)productive even when not engaged in production, since their rooms are there for entertaining, eating, loving, reading, relaxing. The very spaces of reproduction and nurture are hence turned into those of creativity, production and venture. Many makers who do have spare rooms to use for work find their work spilling over into other spaces, seemingly taking over the whole house, and they end up apologising that their home is now a factory within which an improvised living space has to be found.

Just as production muddies and flows into home space, so too childcare flows (or tumbles) into productive space, in perhaps a more uncontrollable fashion. Creating or cordoning off space in which to work and to be creative around the task of childcare presents many makers with a variety of practical problems and tensions when working from home. For one particular maker and single mother, the space in which she found sanctuary away from kith and kin to produce her product was her bedroom, and to create this (un)interruptible space of production she hung this (un)motherly notice on the door, warning: 'At work. Only enter if you're dead or dying!' (Sue, embroiderer). For her, it was the luxury of a (partially) closed door from behind which to work.

For other makers, particularly women but not exclusively so, the kitchen or other downstairs rooms have to be multi-functional. Such downstairs rooms often have spaces within them that are temporarily transformed into productive spaces. In these more indeterminate, temporal spaces, rooms within rooms, where separate rooms are not available, redrawing the boundaries between production and reproduction is necessary. What makers find desirable is an uninterruptible productive space, yet at the same time an interruptible reproductive space. To achieve this, makers redraw boundaries between the two, by cordoning off such spaces with curtains and even bits of string, so creating permeable boundaries. Such membranes between the two filter out certain forms of interaction while allowing others. Pieces of string to mark out the space of production are reportedly among the most effective dividers between production and childcare: an invisible wall, allowing visual and verbal communication between maker and children, while in theory filtering out physical contact. The use of curtains, while effectively reversing the Victorian maxim that children should be seen and not heard, seemed less effective, raising anxiety on both sides of the curtain: 'if they were quiet, I'd think "Oh my God, what are they up to now", and when they were noisy, I had a good idea what they were up to, but couldn't concentrate enough to work' (Sam, marquetry maker).

For Sam, it was a pair of old red velvet curtains which divided off play space for her children from her own work space. While her two young children were still at home, parts of the craft work were transported indoors from the workshop (where the cutting-out processes had been carried out by her partner, Ben) so that she could look after them while making:

> 'the other side of here was another room, behind a curtain, until yesterday; that was where we use to do all the sort of marquetry; I would stick on to the clock and boxes whatever and that was done so I could look after Charlie [her youngest son] and work.'

Drawing the curtains transformed the room into a temporary work space, and Sam completed her work on an otherwise ordinary and quite small kitchen table (around which we sat, joined by her two children, drinking tea during the interview). As Sam's case reveals, such transformations are quite temporary; when the curtains are taken down, the room reverts back to a space more reproductive than productive. Creating such occluded spaces, spaces of production within spaces of reproduction, is not always an effective or plausible form of 'control', of course, since curtains and pieces of strings do not divide off the two spaces concretely. Sam goes on to explain that having young children to look after, despite the curtains, severely disrupted her productive capacities, curtailed her creativity and tended to interrupt the general running of their businesses. Particularly with very young children, the combining of production and childcare in this way presents many stresses and strains:

> 'I think some people can integrate a family.... For myself I'm uncertain ... basically because you can't do anything with them.... It's that kind of limitation and it's also concentration.... So I'm not certain, 'cause I personally find it's just really mind-blowing to try and operate a business with kids, I would rather they weren't there. When you are on the phone I can't bear it if they are telling me that they want to go to the toilet, and just insisting that I take them.... Things always happen when you're on the phone, all hell breaks loose – its uncanny.... It's a complete change of space and having to change to domesticity I find a real ... I couldn't just put Charlie to bed and then go and make a serious [business-related] phone call.'
>
> (Sam, marquetry maker)

Furthermore, when divides are enforced, parents often express feelings of guilt at putting production of their craft before the care of their children. As one basket maker said to me, 'My son never got a look in, I always was finishing off a basket, I was always sat at the kitchen table finishing off another basket; even though I was always there I think he suffered from neglect' (Rachel, willow basket maker). However, for adults as well children, strings, curtains and other strategies have no meaning, and visitors can curtail productive endeavour almost as much as young children, disbelieving that any *real* form of economic production is taking place:

> 'I just can not convince some of my friends that just because I am home does not mean I'm not working. If they call wanting coffee when I am just rushing trying to get an order out, I just have to be rude sometimes and tell them to piss off, it's the only way to get through to them.'
>
> (Jeanette, candle-maker)

PRODUCTION FROM WORKSHOPS

Certain crafts are very dusty, noisy and dangerous; others produce noxious fumes. Crafts such as wood turning may ideally require a separate work space, and yet makers often have to work from home with the divisions of work and home remaining blurred, relatives sitting in among the dust and noise. Where makers do have a workshop, they often tend to have a somewhat stressful love–hate relationship with it. As I will go on to explore in this section of the chapter, as well as workshop spaces evolving from a room in the home, the home can also reabsorb separate workshop spaces. A workshop space is often very costly, both to rent and to heat, some makers paying more to rent their workshops than they do for their homes.

While I was interviewing one particular wood-turner and furniture-maker he pointed to the two upper attic windows of his parents' small farmhouse sited very close to the workshop where he now works (which is no more than a wind-swept barn): 'That's where I started out, in those two rooms up there, and I almost suffocated with the dust and the heat' (Aled, wood-turner/furniture-maker). His separate workshop in the barn evolved from these attic bedrooms. He is now building a second workshop with public display areas, one where he will be dry if not warm all year round. It has taken him well over eight years to develop in this way.

In contrast to this process of the gradual evolution of workshop space where necessary from home space, development agencies are often reported as being keen to 'place' makers in 'proper' spaces of production. Time and time again makers say that they will go along to one of the development agencies for advice on a small loan, for example, and find that the first thing they are offered is a 'purpose-built factory unit' from which to produce. They feel that they are being asked to conform to ideas about serious spaces of production, these factory units supposedly being 'efficient' work spaces. One harp-maker, who refers to such efficient, purpose-built space as 'green sheds', feels that the agencies understand very little about his business of harp-making:

'They wanted to put me into one of their green sheds next to a spot welder on one side and a spray painter on the other; can you imagine... the noise?... There wasn't a window in the place... it may be more economically viable than this mill but you can not reduce us to economic viability; to do so is to deny us our humanity... how could I have made anything there?'

(Arwyn, harp maker)

Makers' space of production can also be an emotional space. The 'type' of space is important; its feel, the lighting, its position, its outlook, as well as 'the cost' are all important to them. So-called purposefully designed productive spaces can, as the makers point out, be found to be the least productive to work in.

However, in not conforming to neat spaces of production, makers undoubtedly feel a certain vulnerability when approached by government agents. They fear that they will not be taken seriously when their spaces of production are 'inspected' to see if they are 'up to scratch'. Aled, the wood-turner and furniture-maker, is involved in a project to build a timber-framed building using local indigenous woods. He told me of his concerns about the visit from the surveyor, remembering that his workshop is now a redundant barn on his family's farm:

'To pass the building regulations the surveyor wanted to come out and see my set up, to see the sawmill and so on. But if he came here, he'd say "NO WAY!!" I don't work from beautiful sheds, and look over here – the muck heap. It's all a jumble. If I did work from beautiful buildings, I doubt there would be any problem.'

(Aled, wood-turner/furniture-maker)

Aled also commented that he felt annoyed that the surveyor should be allowed to see him at work in his totality, in his seemingly chaotic space of production, and he resented the intrusion not only into his work on this particular project but into his future dreams and developments, ideas and products. And so Aled felt that the surveyor would be passing critical judgements on his livelihood, from which 'he is able to jump into his big, flashy car and drive away at the end of the day'.

Where makers do rent workshop space, it often becomes an expensive overhead most makers would prefer to do without, as a cost over which they feel they have no direct control: 'If I could get rid of the overheads of the workshop it would be very nice.... As far as I am concerned... if we could work from home that would be great' (Sally, hand-dyed yarn producer). In order to save money, makers who do have workshops often retreat inside in colder weather – a temporary blurring of the spatial divide. They abandon the workshop seasonally and sometimes daily, because the kitchen by winter and night is preferable to a cold, dark workshop. And as we sat talking, Sally continued to wind skeins of thread. Those, like Sally, who own or rent an outlying workshop would always be bringing some form of work home, if not making directly then doing 'paperwork', as the home is also often the office, a place in which to 'do the books' and to make other arrangements for the business. This can take the form of ordering materials and so on, avoiding interruptive calls through to the workshop or studio and, more importantly, often preventing the costs of having two phone lines, business and private.

There is a continued muddlement of workshop space and home space, and in particular the bringing of work into the home from the workshop and the use of the workshop around the tasks of childcare within partnerships. As Sam and Ben's pattern of working indicated, the workshop space can tend to become 'male' governed, particularly when the male partner is the primary craft maker. Women's roles can then be constructed as more peripheral to the actual processes of production when looking after children, even if women do in effect run the business side of their operation. Jan and her partner Simon produce brightly painted toys: he cuts out the pieces and she paints them up as well as doing all the book work and the business promotion. However, when I first met Jan she said to me, 'Oh, you don't need to speak to me, you need to speak to my husband, I've been looking after the children over the past eight years' (Jan, toymaker). Another maker's wife, Gwenda, sees herself in this support role around her husband's making. She told me about her 'doing the books' in the kitchen and entering the workshop to do 'the monkey work' as far as the business of making was concerned. Yet she also mentioned that she was keeping the business afloat in almost subliminal, unseen ways around the main 'important' tasks of making.

So, as well as redivisioning of spaces, there is also in some sense a recreation of conventional gender divisions, both socially and spatially, as Ken, a stained-glass artist, effectively acknowledges:

Author 'How many are involved in your business?'
Ken 'Just my wife and myself.'
Author 'How does that work out?'
Ken 'We've got a workshop where I do most of the work and everything and I work a lot at home . . . and quite often I bring quite a lot of work home for my wife to do, and in that way she doesn't have to come into the factory; especially if it's a holiday, she can stay at home with the kids and do some work.'
Author 'Does your wife also produce the glass work herself or is she involved in different aspects of the business?'
Ken 'I suppose I'm the boss and I just give her jobs . . . but I think she knows virtually everything in the business, but she has other

jobs with the kids and the home ... she does the bookwork, she does the accounts, she keeps it all up together ... she keeps that side of it.'

(Ken, stained glass artist)

Notice too that his workspace becomes 'the factory', an unsuitable place for women and children.

Where women are the primary craft makers, there tends to be a greater sense of skill-sharing and of the male and female partners taking on an equal role in the business. Unlike Gwenda, who very much played down her role, there can be the placing of the partners on an equal footing. In Justine and Stuart's case, where Justine is the primary maker, Stuart decorates their product and is involved with the selling and business promotion (very much like Jan and Gwenda), but, as this extract from my conversation with him reveals, he comes over very much as 'the heroic new man':

'We look after the children and split the housework between us ... it's not full time at the moment because we have got a 10-month-old baby and a 3-year-old ... so we have brought the business down to a level where we can earn a living from it and cope with the pressures. Having a young family is extremely stressful ... I've got multi-roles, to be adaptable and raise a family and if that means working on the run-up to Christmas till three or four in the morning, so be it.'

(Stuart, flower-grower/paper-making)

Since the arrival of their two children their workshop has been given up, and that is what he means by 'bringing the level of the business down' so that production can once again take place within the home. A sort of reabsorption has taken place.

CONCLUSIONS

The work presented by Peate, Morris and others in the first section of this chapter touched upon the possibilities of craft production within a rural context, and certain of the possible reasons why both had been marginalised within Western capitalist society. Those reasons or discourses are still very evident, and are experienced in the everyday lives of contemporary makers in west Wales. The expectations of the norms of business operation, in particular spaces, developing along particular lines continue to impinge negatively on their operation. As one ex-employee of one of the development agencies in Wales explained,

'We deal with start-ups and they are expected to be on this slope ... which finishes at ICI; the people who start up and employ two people and stay there are failures – they are seen some kind of queer start-up ... and we cannot offer them any support if they do not wish to expand; they are seen as failures even if they have made positive decisions not to develop in the expected way, by employing more people, and taking on management roles and so on.'

(Andrew, ex-development agency)

The household as a base for production was promoted by Thatcherite policies of the 1980s, but there failed to be any deeper political support than the smack of soundbites, as the comments of Weiss make explicit:

the small firm sector works best when competition is restrained by co-operation. As Poire and Sabel observe, in so far as the latter is vital to its success, 'flexible specialisation works by violating one of the assumptions of classical political economy: that the economy is separate from society' (1983: 275) ... this adds up to a world very different from the one envisaged and endorsed by the policies of the British government.

(Weiss 1988: 203)

However, against, and because of, such marginalisation, at a very personal level makers continue to sail in that 'transitory space, between the two unacceptable worlds, between the urban and the rural, production and reproduction, in the ship called Utopia'. As I hope this chapter has shown, it is a contradictory, uneasy and complicated voyage. While they have been 'othered' and indeed choose the position of the 'other', the space they work from can be empowering. In controlling the use of their space, makers play a multiplicity of roles, creatively muddling conventional gendered roles and locations for the performing of these roles. However, makers often have to (re)demarcate spaces of production, 'roping off' work space from play space and creating often permeable, occluded spaces, the result being an ironic reference back to those delimited spaces of production and reproduction that they had previously rejected. Their reweaving of spaces and roles is partly suggested by the title of the chapter. The home as the supposed protective intimate haven becomes muddled with negotiation, competition, machinery, dust, grit, noise and danger. In buying a saw, an ostensibly 'male' tool, the woman maker apparently violates her role as female and sacred mother; she could readily be criticised for spending her maternity benefit on such a dangerous tool of production, since the 'normal' expectations would be that she should have bought nappies, children's toys or clothing. Yet in buying the saw and in attempting to produce from home, she is striving to find a way to look after her children while engaging in production to earn money which can then be spent on nappies and many other items to improve the quality of life.

What is at stake, however, is more than the reweaving of the 'complex geographies of production and reproduction' that forms part of a sub-versive start. I would suggest that focus on their livelihoods gives us a glimpse of these somehow magical liminal spaces where, as Raymond Williams writes, between country and city 'there is a sense of direction' or movement (Williams 1990: 18). This one chapter does not purport to treat the full extent of this movement. What I hope is illustrated here is, perhaps, no more than a series of starts or struggles for individuals, although makers, like Williams, have been known to dream, and to imagine that somewhere in what they are doing there is a sense of some sort of 'new social and natural order' for themselves at least, perhaps momentarily even.

As much as I would like to say that it is a strategy of political paradox, makers living and creating ambiguous spaces, roles and sets of identities which are unfixed and may not be charted, makers at one and the same time celebrate and decry this space of betweenness which they fight to create in order to create from. To be makers they often have little choice but to create or to produce from home, or in fact to buy their first saw with their maternity benefit.

ACKNOWLEDGEMENTS

With thanks to Paul Cloke and Jo Little for editing and commenting upon an earlier, more gawky version of this paper. With special thanks to Chris Philo for never failing to assist generously 'at the drop of a hat', even at distance, and to Laura Williams for helping dreams be realised and con-tinuing to dream. Particular thanks are also due to many others at Lampeter: Eric Laurier, Catherine Nash, Hester Parr and Jill Venus, for their support when I presented a shorter version of this chapter at the Institute of British Geographers, Newcastle upon Tyne, 1995. With thanks also to Rachel Watkins for reading the proofs through.

NOTES

1 A note on reproduction: here I do not merely refer to giving birth to and bringing up children, although households with children do seem to form the focus of the paper. This emphasis on households where children were present was not wholly intended, as I interviewed people living in all sorts of house-holds, married, single, gay, and households where children were absent.

2 'A map of the world which does not include Utopia is not worth glancing at, for it leaves the one country at which Humanity is always landing' (Oscar Wilde).

3 In 1940 H. J. Fleure wrote, 'some of us doubt whether change is always progress and whether change, as it affects us, is not a specialisation in certain directions that cuts off certain possibilities in others'.

4 Weiss (1988) provides a powerful requiem to small family businesses. Under-mining the belief that the neutral and natural progress of business is towards

the large-scale corporation, Weiss ties such developments to blatant nationalism and the state's desire to produce weaponry on a large scale, part of the power of the nation projected through the national or international conglomerate.

5 Peate was concerned that crafts should be adapted to a changing market, rather than being sidelined and lost altogether (see, for example, Peate 1928, 1949).

6 Whatmore provides a useful summary of this condition: '[T]he erosion of the power and influence of rural space and agricultural activity and the refashioning of their structure and meaning in the image of a predominantly urbanised and industrial social order' (Whatmore 1990: 252).

7 Sarah Harper (1989) provides a comprehensive overview here.

8 Makers who moved there often chose to because they had past memories of west Wales, for example visiting the area as a child on holiday or maybe as adults. One couple came on their honeymoon. For others, past family ties were important. One furniture-maker, who lived and worked from his family farm, likened what he was doing to the work of craft makers of the past, providing for the local community, making furniture from locally grown timbers.

9 Makers often made use of abandoned farmhouses, with outbuildings converted for use as a workshop. I have visited a potter working from a cowshed and a harp-maker working from a redundant watermill, with dreams to reinstate the water wheel to power his making from it.

10 This and the other quotations I use in this text are direct quotes, taken from recorded (either taped or hand-written) interviews with makers between 1992 and 1994. All names have been changed to maintain anonymity. For further detail, on the methodology and comments on this technique, I refer interested readers to my forthcoming PhD thesis.

11 Alison Hayford's work (1974) developed within geography the concepts of the gendered divisioning space – work from home, men from women – under capitalism, and in a sense my own work does return to certain questions about 'non'-capitalist economies and gendered spaces.

12 On the domestication of craft production, see, for example, Dalton (1987).

REFERENCES

de Beauvoir, S. (1988) *The Second Sex*, trans. Parshley, H. M., London, Pan Books (First published de Beauvoir, S. (1949) *Le Deuxième Sexe*, Paris, Gallimard.)

Bennett, J. M. (1987) *Women in the Mediaeval Countryside: Gender and the Household in Brigstock before the Plague*, Oxford, Oxford University Press.

Cloke, P. (1989) 'Rural geography and political economy', in Peet R, and Thrift, N. (eds) *New Models in Geography*, vol. 1, London, Unwin Hyman.

—— (1994) '(En)culturing political geography: a life in the day of a "rural geographer"', in Cloke, P., Deol, M., Matless, D., Phillips, M. and Thrift, N. (eds) *Writing the Rural: Five Cultural Geographies*, London, Paul Chapman.

Crouch, D. and Ward, C. (1994) *The Allotment: Its Landscape and Culture*, Nottingham, Mushroom.

Dalton, P. (1987) 'Housewives, leisure crafts and ideology: deskilling in consumer craft', in Elinor, G., Richardson, S., Scott, S., Thomas, A. and Walker, K. (eds) *Women and Craft*, London, Virago.

Droste, M. (1990) *Bauhaus 1919–1933*, Berlin, Benedikt Taschen.

Fleure, H. J. (1940) 'The Celtic West', *Journal of the Royal Society of Arts* 88, October: 882–4.

Gruffudd, P. (1994) 'Back to the land: historiography, rurality and the nation in interwar Wales', *Transactions of the Institute of British Geographers* NS 19: 61–77.

Halfacree, K. H. (1993) 'Locality and social representation: space, discourse and alternative definitions of the rural', *Journal of Rural Studies* 9 (1): 23–37.

Harper, S. (1989), The British rural community: an overview of perspectives', *Journal of Rural Studies* 5 (2): 161–84.

Hayford, A. (1974) 'The geography of woman: an historical introduction', *Antipode* 6 (2): 1–19.

Hoggart, K. (1990) 'Let's do away with rural', *Journal of Rural Studies* 6 (3): 245–57.

James, S. (1991) 'The urban–rural myth or reality', *Geographical Papers*, Reading, University of Reading.

Katz, C. (1992) 'All the world is staged: intellectuals and the project of ethnography', *Environment and Planning D: Society and Space* 10: 495–510.

Knott, C. (1994) *Crafts in the 90s*, London, Crafts Council.

Long, N. (ed.) (1984) *Family and Work in Rural Societies: Perspectives on Nonwage Labour*, London, Tavistock. Cited in Whatmore S., Lowe, P. and Marsden, T. (1991) *Rural Enterprise: Shifting Perspectives in Small Scale Production*, London, David Fulton.

Merchant, C. (1989) *The Death of Nature: Women, Ecology and the Scientific Revolution*, San Francisco, Harper and Row.

Moore, H. (1988) *Feminism and Anthropology*, Cambridge, Polity Press.

Morris, W. (1962) *News from Nowhere and Selected Writings and Designs*, Harmondsworth, Penguin. (First published 1891.)

Naylor, G. (1971) *The Arts and Crafts Movement*, Cambridge, MA, MIT Press.

Pahl, R. E. (1966) 'The rural–urban continuum', *Sociologica Ruralis* 6: 299–327.

Peate, I. C. (1928) 'The social organisation of rural industries', *Welsh Housing and Development Yearbook* 4: 103–5.

—— (1932) 'Welsh folk culture', *Welsh Outlook*, 19 (11): 294–7.

—— (1933) 'The crafts and a national language', *Welsh Housing and Development Yearbook* 4: 75–7.

—— (1943) 'Yr aradaloedd gweledig a'u dyfol', *Y Lenor* 22: 10–18. Kindly translated into English by Cabanski, S. (unpublished).

—— (1949) 'The folk museum', *Journal of the Royal Society of Arts* September: 794–805.

Philo, C. (1992) 'Neglected rural geographies: a review', *Journal of Rural Studies* 8: 193–207.

Poire, M. J. and Sabel, C. F. (1983) 'Italian small business development: lessons for U.S. industrial policy', in J. Zysman and L. Tyson (eds) *American Industry and International Competition*, Ithaca, NY, Cornell University Press.

Said, E. (1978) *Orientalism*, London, Routledge.

Shields, R. (1991) *Places on the Margin: Alternative Cultures of Modernity*, London, Routledge. Cited in Gruffudd, P. (1994) 'Back to the land: historiography, rurality and the nation in interwar Wales', *Transactions of the Institute of British Geographers* NS 19: 61–77.

Shiva, V. (1989) *Staying Alive: Women, Ecology and Development*, London, Zed Publications.

Shurmer-Smith, P. (1992) 'Cixous' spaces', paper presented to Department of Geography, University of Wales, Lampeter. See also by the same author (1994) ' "Cixous' spaces": sensuous spaces in women's writing', *Ecumene* 1 (4): 349–362.

Weiss, L. (1988) *Creating Capitalism: The State and Small Business since 1945*, Oxford, Blackwell.

Whatmore, S. (1990) 'Theories and practice for rural sociology in a new Europe', *Sociologica Ruralis* 30–31 (3/4): 251–9.

—— (1991) *Farming Women: Gender Work and Family Enterprise*, London, Macmillan.

Whatmore, S., Lowe, P. and Marsden, T. (1991) 'Artisan or entrepreneur? Refashioning rural production', in Whatmore, S., Lowe, P. and Marsden, T. (eds) *Rural Enterprise: Shifting Perspectives in Small Scale Production*, London, David Fulton.

Williams, R. (1980) 'Between country and city', in Pugh, S. (ed.) *Reading Landscape: Country, City, Capital*, Manchester, Manchester University Press.

14

POOR COUNTRY
Marginalisation, poverty and rurality
Paul Cloke

OTHERNESS AND RURAL POVERTY

He lives in a house, a very big house in the country
Watching afternoon repeats and the food he eats in the country
He takes all manner of pills and piles up analyst bills in the
country
It's like an animal farm, lots of rural charm in the country

Blur, Country House (1995)

As has been evident throughout this book, constructions of rurality in both geographical and cultural arenas have tended to map out maps and paint pictures in which social problems have been (if you will pardon the pun) blurred. In the fertile imagination of a band whose penchant is to reveal ordinary life 'as it is', the apparent idyll and charm of a rural setting is mapped against the reality of anxiety, boredom and corruption by power. Yet the focus here is the 'man' in the 'very big house' who used to be a successful city dweller but who escaped the rat race only to find that country life is not all that it is cracked up to be. As in *Animal Farm*, an escape from the old (urban?) tyranny is merely replaced by a new (rural?) tyranny. Thus the supposedly problem-free nature of rural life is exposed for the myth that it is, but the 'victim' in the spotlight remains assuredly in the mainstream of 1990s cultural values and socio-economic positions.

In this chapter, I want to touch on a number of questions about the 'otherness' of the 'poor' and 'poverty' in rural societies. These questions relate principally to the implications of a cultural turn in social science for the continuing identification of, and protest against, the 'evil' of poverty. Given that otherness is usually encountered in relation to selfness, I should immediately indicate my own moral biases here. My own geographical imagination is fired by social issues such as poverty at least in part because of a moral rootedness in Christian thought and socialist Christian ideologies (see Cloke 1994 for a fuller explanation of this), which identify poverty as both (literally) evil and a crucial target for political action. I have, therefore, a keen interest in the material circumstances of individuals and households as well as in the power relations associated with the powerlessness of poverty. Similar concerns, of course, are traceable to other moral roots in other people. However, the 'morality' of geographies of poverty also involves a more Kantian concern to understand the moral suppositions which inform the lives of particular people in particular places. Driver (1988) has been prominent among those who have argued that 'outside' researchers need in some senses to unlearn their own moral presuppositions and instead to seek out those of their research 'subjects' in order to evaluate and interpret their findings. 'Giving voice' to marginalised people involves a restraining of the researcher's voice as well as an amplifying of the voice(s) of the researched.

It is important to question whether these different cuts on morality – the personal imagination of the researcher and the revealed presuppositions of the researched – can coalesce without mutual exclusion. Recent debates on the potential for bringing together political economy and cultural studies help us here. Sayer's (1994) concerns for the neglect of political economy in the emphasis on all things cultural may be contrasted with Barnes' (1995) search for symbiosis in recognising the importance of the cultural within the political economic. More specifically, Leyshon's (1995) fears for what he sees as the increasing marginalisation of poverty as a topic of study may be contrasted with McCormick and Philo's (1995) wish to tease out the hidden geography of poverty in order to reveal poverty in 'other' forgotten places (see also Philo 1995). Are we really faced with the choice between a normative and insensitive imposition of the idea 'poverty' onto pre-formulated social groups clearly located by pre-theorised processes of inequality, and a sometimes touristic (and maybe even voyeuristic) intrusion into the different experiences of what might be called poverty in some individual discourses but which is often hidden from sight by the broader exercise of discursive power? Each 'option' caricatures the other: the underconstructed and transcendentalist categorisation of poverty meets the conviction-less identity politics of poverty as just another 'other'. My account of those positions is too simple, yet I am struck by the very real fears being expressed about 'missing the point' in our studies of poverty.

One way forward might be to encourage some further (en)culturing of political economy approaches. As a starting point we can identify the twin foci of material and ideological inputs to the 'reality' of poverty:

> we must continue to recognise empirically (and to continue theorising) the processes which discriminate between different categories of individual in the apportionment of material 'rewards' and 'opportunities' ... but perhaps we now need to supplement this material focus with a greater sensitivity to the complex ideological

inputs to the reality of social differentiation, and in so doing ask questions about the ideologies – and at this point it is appropriate to speak of the blatant moral judgements which serve to 'naturalise' ... various forms of social differentiation.

(Philo, 1991, 18–19)

That this is only a starting point is evidenced in the recent debate between Philo (1992, 1993) and Murdoch and Pratt (1993, 1994) in *Journal of Rural Studies*, in which broad agreement over material and ideological inputs to understanding 'problems' in rural societies led to some dispute over whether the *purpose* of such research was to help change the world or merely to give voice to difference. For myself, poverty is not just another other, so I retain a sense (however idealistic) of wanting to contribute to doing something about it. In seeking to sharpen the focus on the blurred problematics in rural life it is insufficient merely to gaze on the 'man' (or woman) in the 'very big house' (or without a home) in the country. In the remainder of this chapter I present a brief account of studies of poverty in rural England and (at a less detailed level) in the rural USA with which I have been involved. In each case (although differently), discursive codes, symbols and concepts of poverty and welfare have differentiated spatially between the urban and the rural, thus mediating the political significance attributed to the issue of poverty in rural areas.

POVERTY IN RURAL ENGLAND

Cultural constructions of (non-)poverty

In October 1995, the UK government produced a White Paper entitled *Rural England: A Nation Committed to a Living Countryside* in order to open up a debate about the future of rural areas. Of the 146 pages of discussion, the nearest the White Paper comes to addressing the issue of rural poverty is as follows:

> Yet the attractiveness of the countryside as a place to live can easily disguise the fact that rural society faces its full share of problems. Rates of pay, especially in traditional rural occupations such as agriculture and tourism, are generally below those in urban areas, a disadvantage often compounded by the seasonability of the work. Although unemployment is below the national average, there are nonetheless problems associated with underemployment and low income, often cheek-by-jowl with relative affluence. These problems may be less visible and more dispersed than in the town but rural deprivation brings its own disadvantages, particularly in terms of access to services.

(p. 62)

One year previously, the Rural Development Commission – a quasi-governmental agency within the auspices of the Department of the Environment – published a report entitled *Lifestyles in Rural England* (Cloke *et al.* 1994), which suggested that the proportion of households in twelve study areas in rural England who were 'in or on the margins of poverty' was as high as 39.2 per cent in one area, and over 20 per cent in nine of the twelve areas. Why, then, did the White Paper eschew discourses of poverty in a rural setting, preferring terms such as 'disadvantage' and 'deprivation' (and then only fleetingly)?

At least part of the answer lies in the 'othering' of poverty within dominant social and cultural constructions of rurality and rural life. The broad political context makes such othering very comfortable in 1980s and 1990s Britain. In 1989, the then Secretary of State for Social Security, John Moore, reported that economic success had ended *absolute* poverty, and that *relative* poverty was a misguided notion that simply meant 'inequality' (see Oppenheim 1990). However, despite the wholesale censorship of the word 'poverty' in government documentation, there is a strong sense in which it has been easier to deny the existence of the poor in rural areas than in urban. Several authors (see Cloke *et al.* 1995a for a summary) have interconnected the imagined geographies of idyll-ised rural lifestyles with the idea that poverty in rural areas is being hidden or rejected in a cultural dimension both by rural dwellers (including those who appear, normatively, to be 'poor') and by decision-makers with power over rural policy. And here, the metaphor 'deprivation' has proved most useful to those who will not use, or do not even think of using, the notion of poverty in a rural context. Deprivation presents a word-bin into which all manner of incongruities in rural life may be placed, without any particular impetus to ascribe them as problematic, or to do something about the problems concerned. Thus, rural people can be recognised as 'deprived' of ready access to the advantages of urban life, but need not be thought of as impoverished by such deprivation because of the perceived compensations which are a vital construction of rural living. In this way

> the poor can thus be disregarded as being 'content' with their (undemanding) rural life, and the not-so-poor will not be able to reconcile the idea of poverty with the idyll-ised imagined geographies of the village, so any material evidence of poverty will be screened out culturally.
>
> (Cloke *et al.* 1995a: 354)

In their study of poverty in rural Leicestershire, Fabes *et al.* (1983) have suggested two main ways in which the mythology of imagined rural idylls tends to direct attitudes towards rural poverty. First, they point out that it is the very 'deprivation' of rural areas in relation to cities which makes them so attractive to urban in-migrants. It is the *lack of* development, transportation and general opportunity which reproduces rural space as an object of desire – as marginal to the modernist tendencies of the city, and a safe haven from the dysfunctionalities inherent in the cramming together of 'decent' middle-class citizens with 'other' kinds of people. Secondly, the idyll-ised cultural construction of rurality serves to conceal the poverty which does co-exist with the affluence which glosses the representative nature of rural spaces.

> In this respect, the poor unwittingly conspire with the more affluent to hide their own poverty by denying its existence. Those values which are at the heart of the rural idyll result in the poor tolerating their material deprivation because of the priority given to those symbols of the rural idyll: the family, the work ethic, and good health. And when material deprivation becomes so chronic by the standard of the area that it has to be recognised by the poor them-selves, shame forces secrecy and the management of that poverty within the smallest possible framework.
>
> (Fabes *et al.* 1983: 55–6)

In this way, it is possible to present a very cogent case for the discursive power of rurality in signifying itself as a poverty-free zone. Constructs of rural idyll(s) concomitantly *exacerbate* and *hide* poverty in rural geographic space. Moreover, these very idylls permit a political interpretation that these are problem-free areas. Accordingly, rural areas have become a repository of late twentieth-century English values, demonstrating spatially the virtues of a privatised and deregulated national political economy. Rurality represents a non-interventionist arena where late capitalist markets have permitted idyll-ised lives, and therefore policies associated with the rural arena have been muted to fit the discursive 'filtering out' of problems.

However, the messiness of poverty in rural life comes sharply into focus when we examine the implications of restraining the normative instinct in our researchers' voices and relying on lay discourses of rural poverty, for, as the study by Fabes *et al.* makes clear, there is an extent to which merely to give voice to the rural poor may not deconstruct fully the poverty-free representations of rural life. Material poverty does not speak for itself. So, just as it seems inappropriate merely to impose transcendental normative categories onto rural people in order to show them to be 'poor', it seems equally inappropriate not to question the contexts in which the rural 'poor' will often deny their own 'poverty' in their lay discourses of rurality. Indeed, it could be that the rural poor are disenfranchised by the language of their own poverty. Moreover, the reactions of rural people towards their own poverty tend to be differentiated rather than uniform. The study by Fabes *et al.*, for example, identified several different responses to a seeming position of poverty: a fatalistic acceptance of a lowly position within a clearly demarcated social hierarchy, a high tolerance of poverty, with people 'scraping by' for as long as possible in their current positions; a distinct lack of material aspirations, with priority being allocated instead to family life, good health, and the richness of the rural cultural heritage; and a stigmatic burden of shame and secrecy which then militated against the seeking of help. In each of these cases, lay discourses may be interpreted by some as confirming rather than contesting countryside cultures of problem-free idyll-ised life. In order, then, not to 'miss the point' about rural poverty it seems preferable to focus on both the structuring of opportunities and the different experiences of those opportunity structures. This dual focus may be illustrated empirically by the findings of the *Lifestyles in Rural England* study mentioned above.

Marginalisation and poverty in rural lifestyles

Important new evidence about rural poverty in contemporary Britain has appeared during the 1990s. For example, studies in Scotland (Shucksmith *et al.* 1995) and Wales (Cloke *et al.* 1995b) have provided very clear indications of both the scope and the scale of difficulties relating to low-income living in the countryside. Although, as always, precise normative definitions of 'poverty' can be (and are) disputed, the illusory representation of rural Britain as an affluent and problem-free series of spaces is being severely challenged by accounts of the living conditions, opportunity levels and life experiences of rural people in these studies.

Perhaps the most widescale study of rural life within contemporary Britain has been the *Lifestyles in Rural England* research, which interviewed 250 households in each of twelve case study areas (see Cloke *et al.* 1994, 1995a, 1995c, 1997). It is important to emphasise that such a study permits many different windows onto issues of poverty and social marginalisation (see, for example, Woodward 1996), but the major findings do provide grounds on which to propose the continuing significance of 'problematic' rural lifestyles (Cloke 1993). The Rural Lifestyles study confirmed not only that very evident problems of low income and poverty remain in rural England, but also that rural people tend to associate the power of income with the power of status in their reflections on the processes of social marginalisation in England's villages. As Cloke *et al.* (1997) point out, 'the "problematic" associated with poverty ... is vitally interconnected with experiences of powerlessness and marginalisation drawn from living cheek by jowl with affluent people of status and power' (forthcoming). The twelve case study areas surveyed by the Rural Lifestyles study ranged from what might be generalised as relatively 'low-income areas' (for example, North Yorkshire and Northumberland) to 'high-income areas', such as Warwickshire and West Sussex. One of the major aims of the study was to investigate the issue of low income and poverty in the areas concerned, and, drawing on the previous work of Bradley *et*

Table 14.1 Households in or on the margins of poverty definitions: study areas ranked for each of three indicators

Mean*		Median**		Income support***	
North Yorkshire	61.9	Devon	47.1	Nottinghamshire	39.2
Northumberland	61.1	West Sussex	45.8	Devon	34.4
Nottinghamshire	58.8	Essex	44.4	Essex	29.5
Shropshire	51.4	Suffolk	42.6	Suffolk	25.5
Devon	50.0	Nottinghamshire	41.2	Wiltshire	25.4
Wiltshire	49.3	North Yorkshire	40.5	Warwickshire	22.6
Warwickshire	48.4	Warwickshire	38.7	North Yorkshire	22.0
Cheshire	46.2	Northumberland	37.5	Shropshire	21.6
West Sussex	45.8	Wiltshire	36.6	Northamptonshire	14.8
Suffolk	44.7	Shropshire	35.1	Cheshire	12.8
Across 12 areas	51.2		40.6		23.4

Source: Cloke *et al.* 1994
Notes:
* In or on the margins of poverty: < 80 per cent mean
** In or on the margins of poverty: < 80 per cent median
*** In or on the margins of poverty: < 140 per cent income supplement entitlement

al. (1986) and McLaughlin (1985, 1986), the study utilised three normative indicators of relative poverty (Table 14.1). The first two indicators use data on mean and media incomes, while the third measures poverty in relation to state benefits (Townsend 1975, 1979).Townsend's indicator is perhaps the most rigorous of the three, and it identifies significant spatial variation in the proportions of households considered as 'in or on the margins of poverty'. In some case study areas – Nottinghamshire, 39.2 per cent; Devon, 34.4 per cent; and Essex, 29.5 per cent – there would seem to be very severe levels of income-related problems among rural people. Elsewhere – West Sussex, 6.4 per cent; Cheshire, 12.8 per cent; Northamptonshire, 14.8 per cent – the aggregate problem appears to be far less severe.

Three cautionary conclusions are drawn from these normative indicators (Cloke *et al.* 1994: 95):

1 The fact that nine out of twelve study areas contained 20 per cent or more households in or close to poverty itself suggests a severe problem for rural lifestyles in most areas of the country. Neither should we in any way discount the severity of problems experienced by the 14.8 per cent of households in Northamptonshire, the 12.8 per cent in Cheshire and the 6.4 per cent in West Sussex who suffered the additional burden of being a smaller minority with income problems in areas of greater affluence.

2 There appears to be a mix of rural geographies at work here. Although the more urbanised study areas tended towards having a lower pro-

Table 14.2 Percentage of households moving into the area over the past five years with incomes under £8,000

Area	Percentage
Northumberland	38.8
Nottinghamshire	35.6
North Yorkshire	35.4
Devon	22.5
Suffolk	20.0
Essex	17.0
Northamptonshire	16.2
West Sussex	16.1
Shropshire	15.7
Cheshire	9.8
Warwickshire	8.5
Wiltshire	7.9

Source: Cloke *et al.* 1994

portion of households categorised by this index, and the more remote areas tended to have a higher proportion, the cases of Essex, Suffolk, Wiltshire, Warwickshire, North Yorkshire and Shropshire indicate that such tendencies were subject to alteration by local circumstances. We need to look beyond all-encompassing structural factors or broad concepts such as 'remoteness' and 'pressure' in the understanding of these phenomena.

3 It has been argued that the issue of poverty and deprivation is an outdated phenomenon, becoming anachronistic in the 'prosperity' of the 1980s. Our findings suggest that the issue not only is very important in our 1990 surveys but also is being reproduced by patterns of in-migration.

This last finding reflects the fact that those case study areas where aggregate incomes were lowest, and poverty indicators were most significant, figured strongly as reception of lower-income in-migrant households (Table 14.2). Surprisingly, however, a number of other case study areas were also receiving significant proportions of lower-income in-migrants, with only Cheshire, Warwickshire and Wiltshire – the most gentrified and protected housing markets of all the study areas – receiving fewer than around 15 per cent of households with incomes below £8,000. Such movements may partly have been connected with people moving in to retire on low income but higher savings. However, there is a suggestion here that the broad problems associated with low income and poverty are being reproduced more generally as well as in the more obvious 'low-income areas'.

In one sense, these normative accounts merely offer a preconstructed and potentially 'othering' process of researchers finding what they were looking for and at the same time informing rural people of a 'poverty'

259

which may not be acceptable to the people concerned. I would argue that normative contextual material such as this is a vital ingredient in our understanding of the material nature of (lack of) rewards and (lack of) opportunities. The Rural Lifestyles study confirmed that processes associated with housing, job and service markets were discriminating against and between different categories of individuals. It is these processes, and the power relations which underlie them, which are capable of marginalising particular households and individuals – setting them apart as others to the accepted characteristics of the rural self relating to affluence, contentment, mutual co-operation and shared experience. In terms of opportunity structures at least, there seems to be a diversified rather than unified distribution of opportunity-based experiences in these rural areas.

However, the *experience* of rural life is clearly not solely explainable in terms of the distribution of income and other opportunities. Accordingly, the contextual recognition of differentiating processes of opportunity structure does need to be supplemented with a sensitivity to the complex culturally constructed inputs to the experienced 'reality' of socially differentiated rural life. In other words, within similarly structured opportunities there will be different life experiences, variously impacted on by the naturalising moralities which include or exclude those who do and do not 'belong' to the country(side). In Bell's (1994) study of Childerley, a small village outside London, he speaks of the material conditioning the cultural:

> As in communities everywhere, material circumstances condition the cultural fabric of Childerley, resulting in contrasting habits of feeling across the class divide. The back-door spirit of working-class villagers I described as more informal, group-oriented, interactive, local and experiential. The front-door spirit of the moneyed people I characterised as more formal, individualistic, private, far-flung and distanced.
>
> (Bell 1994: 163)

This recognition of different worlds within the rural habitus is important, and offers scope for viewing 'back-door spirit' as a potential space of resistance by which the oppressed can inhabit different cultural territory from their oppressors. There are, however, other connections between the material and the cultural, most notably the ways in which dominant cultural constructions of rurality can subsume the potentially problematic and marginalising experiences of rural life into a seemingly hegemonic 'take-over' of what meanings should be attached to that rural life. Rather than the material conditioning the cultural, we can suggest a series of complex, interconnecting and dynamic relations between the material and the cultural in which, for example, the rural poor themselves become sucked into versions of rural life which are not supported by their life experiences.

The Rural Lifestyles study did not employ the ethnographies necessary to illustrate fully these complexities (see Cloke 1996). Nevertheless, some of the comments made by the people interviewed do serve as pointers to the cultural cloaking of material circumstances. For example, discourses of 'deprivation' were clearly constructed in terms of wider discourses of rurality, such that problems associated with deprivation were hidden from many rural residents:

'I don't see any deprivation, but I am sure there are deprived people in all areas.' (Northumberland)

'Deprivation? I don't know what you mean. We all help each other.' (Warwickshire)

The hidden nature of deprivation was clearly associated with the ways in which 'rural life' was being constructed differently from town life:

'There is no deprivation – if you're used to living in the town you think differently to people living in the country and vice versa.' (Wiltshire)

'People in the countryside live their lives in different ways.' (Suffolk)

'If there is any deprivation I suppose that those people that live in the country have more simple ideas and don't feel deprived.' (Northumberland)

'It seems that people have lower expectations in rural areas, so they put up with it.' (Shropshire)

Here, 'people' in the country are clearly not being differentiated. It is as if there is a unified moral code of 'expectations' which is given a low common denominator in terms of a 'different' and 'simple' life with potentially 'lower expectations'. These are, however, cultural constructions of rurality which favour the adventitious, the affluent and the mobile, whose enjoyment of the rural object of desire is not diminished by having to accept a 'simple' lifestyle based on 'lower expectations'. Their role in this idyll-ised culture is assured, and others find it necessary to subscribe to the idyll for fear of further exposing their own marginalisation from it. Anyone not conforming to the naturalised problem-freedom can be explained away in terms of intersections between local rural relations and the wider (non-rural) relations by which an individualist politics allows the blaming of the victim. Some of our respondents therefore saw the 'undeserving poor' as failing in their duties of self-help:

'People should try to help themselves to some extent; some people just sit back.' (Northumberland)

'Deprivation? There is none – it's up to the people who live here to make the most of their lives as we all do.' (Shropshire)

'Very rough people seem to get houses around here now, unmarried mothers, etc.' (Nottinghamshire)

The conclusions of the Rural Lifestyles study suggest, therefore, that some of the differences in the lives of rural people are affected by a 'structuring from without', but that differences also occur in terms of how lifestyles are 'experienced from within'. Some diversity relates to the different contexts and situations people find themselves in. It is also the case, however, that cultural assumptions about rurality appear to be important in the differential constructions of rural lifestyles. Newer in-migrant residents often carry with them a rather sanitised version of idyll-ised rural life. Their recognition of the problems experienced by 'others' depends at least in part on whether their hegemonic cultural assumptions about rurality permit them a view not only on the need for tangible opportunities (housing, jobs, etc.) in their rural homeland, but also of the less tangible need for others who are economically, socially or culturally different to belong to that rural homeland.

POVERTY IN RURAL AMERICA

The rural 'deserving' poor

In 1993, a task force of the US Rural Sociological Society reported its findings on the problem of 'persistent rural poverty', and concluded:

> The national government has done little that really matters in eliminating the causes of rural poverty. At most, US policymakers create marginal relief, disguised behind a thin veneer of political rhetoric. . . . Market problems and social failures in rural America are different and in many ways more severe than in metropolitan regions.'
> (Rural Sociological Society Task Force 1993: 292)

A brief review of studies of poverty in rural America provides a significant contrast with British studies in a number of respects. To begin with, the notion of rurality as cultural idyll does not translate comfortably into the American context. Although there are some rural areas, close to metropolitan centres, which attract in-migration on the grounds of culturally ideal-ised landscapes and communities (areas of New England provide an example here), rurality in America is diverse, complex and not generally conducive to the ethnocentric ideas about idyll which have captured the gaze of British studies. Many parts of rural America are not constructed as objects of desire and landscapes of pleasurable historicity. Rather, rural areas are variously imbued with meanings associated with pioneerism, wilderness, the-back-of-beyond and backwardness (see Lapping 1992). Regional cultures of the rural are perhaps more useful than national-scale constructs in this case (Cloke and Milbourne 1992).

Rural America is also very different from the foregoing account of Britain because it is subject to an official state-defined poverty statistic. The poverty line was developed in the 1960s and was based on less than generous assumptions about necessary budgets for food and essential household items. The measure has been heavily criticised (Ruggles 1990), and 'most experts consider the line is set too low' (Rich 1989). Nevertheless, the existence of a poverty line means that the state finds it very difficult to employ discursive rhetoric about there being no poverty. Rather, the emphasis has been on the appropriation of particular social constructs of poverty in order to differentiate between the deserving and undeserving poor (Katz 1989; Cloke 1992). The prevalence of poverty in non-metropolitan rather than city spaces is counter-intuitive to most Americans. This is at least in part due to the popular construction of urban underclasses as undeserving, and the non-metropolitan low-income workers as deserving. The glare of publicity falls on the former rather than the latter, and so the cultural construction of poverty in America thereby has a distinct spatial outworking, with impoverished rural Americans being 'lost in the shadows' (McCormick 1988: 21). Even when poverty is commonly recognised in rural areas, the tendency is to assume that it is focused on particular problem regions, notably Appalachia and the South. However, 'Today, the problem has no boundaries. A tour of America's Third World can move from a country seat in Kansas to seaside Delaware, from booming Florida to seemingly idyllic Wisconsin' (ibid.: 22).

The Rural Sociological Society Task Force (1993: 78) outlines seven major trends in American rural poverty:

1 The gains in reducing poverty made during the late 1960s and 1970s were largely lost during the 1980s. The levels of rural poverty in the late 1980s were almost 20 per cent.
2 The poverty rate has been consistently highest in non-metropolitan areas; poverty is proportionately a larger problem in rural America than in urban places.
3 The rural–urban gap has been getting larger over the past fifteen years.
4 The rapid growth in poverty during the 1980s is not explained by the rise in female-headed families.
5 The rural poverty rate is more sensitive to unemployment than is the rate of urban poverty.
6 There is a 'suburban ridge' in the spatial distribution of income. It is as though central cities are poverty craters surrounded by a ridge of high income beyond which lies a plain of poverty reacting to the next suburban ridge.
7 Black rural poverty is extremely concentrated in the South: 97 per cent of rural blacks with incomes below the poverty threshold live in the South.

This evidence for the importance of rural poverty in the USA is supplemented by accounts of the anatomy of the rural poor (see Barancik 1989; Gorham and Harrison 1990; Porter 1989; Tickamyer and Duncan 1990) which confirm that rural poverty is characterised by predominantly white, two-adult households which include at least one paid worker. By implication, rural areas are not viewed as locales where the most important and qualitatively 'worst' forms of poverty exist. Rather, they represent the acceptable symbols of poverty, where 'hard times' are somehow naturalised into farming landscapes, and where the cultural signification of good honest countryfolk somehow seems to purge rural communities of the key poverty characteristics of social isolation, dysfunctionality and welfare dependency. The cultural myth here is not, then, one of idyll-istic lifestyles which are problem-free, but rather one of pioneer lifestyles, where hardship is an integral part of the 'joy' of taming nature and extending the boundary of civilisation. As Porter (1989: 21) states, 'the vast majority of the rural poor do not fit the common stereotypes of poor'. Yet as in Britain, these myths merely cloak life experiences which cannot be naturalised away as some kind of inevitable rural inheritance. Consider the story of Ida Swalley from Ottawa, Kansas (population 11,000):

> 'Poverty is passed from one generation to another: it is the only legacy of the poor. Ida Swalley married at 15 to escape a hard-drinking stepfather. She has no marketable skills. Now 43, she is separated from her fourth husband and is living in a squalid $200-a-month apartment that could be owned by an urban slumlord. Swalley shares the hovel with her 17 year old son and a menagerie of bugs and mice. An old fly swatter is the sole decoration on one wall. The Kansas heat pushes the fetid air towards 100 degrees and aggravates Swalley's heart problems. She says things may improve once her new boyfriend gets out of jail. Her fondest hope is that life will somehow be better for her daughter, Carol Sue, 26, and her two-year-old granddaughter, Jacqueline Ruth. But that dream may be illusory. Carol Sue Stevens earns just $3.85 an hour as a nursing-home aide. Her life, like her mother's, has been a succession of small-town romances with men prone to drunkenness and violence. Little Jacqueline Ruth was fathered by Carol Sue's current boyfriend, but the toddler doesn't carry either parent's surname ... "If we end up in some custody fight, I don't want her in court already using her daddy's last name".'
>
> (McCormick 1988: 22)

The irony of the 'deserving' poor such as Ida and Carol Sue is that the largely individualist politics of welfare means that their very perceived ability to fend for themselves (a key characteristic of the 'working poor') is a disincentive to further state support. In one of America's top wheat-producing states, their share of nature's bounty which surrounds them is symbolised by standing in line for government food parcels of flour.

Regional identity and poor country: Appalachia

The interconnection of material apportionment of opportunities and more cultural experiences of social marginalisation is more clearly visible on the local, rather than national, stage in the USA. Perhaps the most notorious region of rural poverty in the USA is Appalachia (Batteau 1983; Billings 1974; Lewis *et al.* 1978; Walls and Stephenson 1972). In the 1960s, the Johnson administration categorised Appalachia as a region of 'grinding poverty,' and so the region became a 'frontier in the war on poverty' (Precourt 1983: 86), where material needs and cultural values and beliefs became intertwined in the ideology of poverty. As in the case of some rural 'poor' people in Britain, people in Appalachia are generally not disposed to accept the stigmatic tag of 'poor' or 'deprived'. As Stephenson (1968: viii) remarks,

> I, like many people raised near the Appalachians, was not so aware that we had such problems until someone informed me.... Only gradually did I come to realize that the people referred to by Michael Harrington [1962], Harry Caudill [1963] and John F. Kennedy and Vance and the *Saturday Evening Post* – were the same ones I had as neighbors and school friends when I was a child. In truth, I still think of the mountains as a corner of heaven first and a national disgrace second.

Appalachia's poverty has bound people to region in a complex inter-relationship of opportunity and lifestyle. The resultant identity both characterises the region and is characteristic of it. Part of the identity relates to a history of political-economic development, land ownership, resource extraction (notably coal-mining) and labour relations. Tickamyer and Tickamyer (1987) describe a historical narrative of exploitation and external control of land and labour which has created regional underdevelopment and an economy which often works to the detriment of its residents. High unemployment and poverty reflect the weakness of labour and the direct result of exploitative social relations between the small minority who have benefited from this form of economic development, and the rest of the population.

On the other hand, Appalachia's regional identity also draws on a unique regional subculture, with its own patterns of economic activity, family life, language and customs which run counter to modern economic values. In this subculture, residents appear to 'accept a poverty-level lifestyle in return for the supportive environment provided by community and kin networks' (Tickamyer and Tickamyer 1987: 15).

The narrative on Appalachian subculture is partly from the outside looking in. Lohmann (1990) points to pop-culture stereotypes in America, such as Lil Abner and Snuffy Smith, which have represented the region in terms of an acceptance of subsistence lifestyles and high levels of poverty, often seen as a region settled by rugged individualists ('mountaineers') who are interested only in their private little worlds:

> Thus, the mountaineer appears to be at variance with the standardized image of the American in everyday life. Consequently, he is accused of possessing negative attitudes, of being a defeatist, of having an inferiority complex, and of lacking appreciation for education. His lack of social skills in modern social institutions is dubbed by some as having a 'backwoods flavor'. His inability to follow expected behavior patterns in group situations is assigned to what some call 'rural values'.
>
> (Zeller and Miller 1969: 11)

Shapiro (1978) dismisses this 'myth' of Appalachia as largely a fabrication of journalists and intellectuals which has been supplemented by missionaries and 'local color writers' who fostered (among other things) the arts and crafts movement in the region. However, there is a sense in which something of a similar subcultural narrative is told in discourses from the inside looking out. Lohmann (1990) tells of a unique and cherished cultural heritage which has been both encouraged and promoted by Appalachian people themselves. This narrative is at odds with everyday Americanism, but rather than agreeing to the masculinist 'culture of poverty' outlined by Zeller and Miller, the insider story is one of independence, and of learned knowledges relating to coping strategies for experiences of poverty. In Dana Wildsmith's poem 'New Poor' we see a feminised experience of poverty: a woman coping, learning, employing old knowledges, knowing new poverties; a view from the inside looking out:

> First she finds out how much time it takes to be poor.
> The path from some to less has to be planned,
> even meals: each must bleed to the next,
> the whole roast to the hash,
> yeast bread to toast,
> toast with eggs.
>
> This is a siege. She sits down to take stock, to deal
> not with what is gone, but with what is left.
> Each thing she names, she owns as a tool:
> milk and eggs can be two meals or
> cake once. There is still tea.
> In siege time,

old rules flesh out like a starch-fed child: girls were
once taught to save back a bit to use when need came
(from a quart, the scant up; from four eggs;
a yolk), to let bread pad out poor meat,
and that a whole, sliced thin, seems
more than it is.

Meals laid on the best plates take on a bit of grace.

When the siege was new, she mapped these hints as if
they were clues to a code, as if one day at last
each dear act of faith would add up to
some truth less stark than the one
she now knows: that there are
no tricks to use on
food that has run out.

(Wildsmith 1990)

It is perhaps easy to see how indigenous cultural identity chimes with external caricature of popular representation. It is also perhaps easy to miss the point about poverty, as Leyshon has suggested. Cultural constructions of Appalachian rural identity mythologise and even naturalise some of the opportunity privations experienced by rural people, but these very experiences are reflected upon in the contact of both internalised and externalised constructions of rural life. Thus the material and the cultural appear inextricably intertwined, with othering by outside selves and self-othering by Appalachian people, going hand in hand.

OTHER POVERTIES?

If Blur's 'country house' is as close as many people get to a knowledge of the problematics of rural life, then there is an important role for researchers in seeking to sharpen the focus on these blurred problematics. Such a task will inevitably include attempts to understand the moral suppositions as well as the material opportunities which inform, enable and constrain the lives of particular people in particular rural places. The contrast between rural England and Appalachian America is immense. In rural England, the overriding cultural logo is one of problem-freedom. Poverty is 'othered' within dominant social and cultural constructs of rurality and rural life. The othering of poverty acts as a cloak, which keeps material evidences and experiences well hidden – so much so that the very existence of poverty can be contested ideologically. To some extent 'poor' rural people will go along with this othering process, perhaps reflexively seeking to place themselves in whatever way possible as belonging to a hegemonic idyll-ised rurality. In the rural USA, rurality is signified by problem-fullness rather than problem-freedom. Poverty is accepted as an often naturalised facet of the pioneering and even backward life 'out there'. Where 'othering' occurs, it seems that in some senses whole regional identities can be offered in terms of the mainstream social and cultural constructions of American life. The othering of regional identities seems to explain away the material evidences and experiences of poverty – the 'deserving' poor can thus be packaged up neatly as experimental laboratories for regional planning, or as demonstrating considerable potential for the theatrical spectacle of rural life demanded by growing tourist industries (see Wilson 1992).

Two concluding thoughts occur to me here. First, it is clearly very important for researcher's to continue with the task of giving voice to people experiencing poverty. Yet, even if it is possible to exercise appropriate constraints on the researcher's own voice, it seems likely that the voices of the rural poor will not tell an unambiguous story. There are so many reasons why the poor will reflect in particular ways on their own poverty – sometimes to deny it; sometimes to explain it away; sometimes to be frank about problematic experiences, sometimes not. Clearly, the researcher's voice is not the only potential source of external moral supposition to be restrained in order to hear and understand the voice of the subject. Lay discourse is enmeshed with all kinds of reflexivities and coated with innumerable external discursive references. Restraining these references and associated moralities and codes of meaning is a task not to be underestimated.

Secondly, this chapter has not begun to deconstruct the seemingly monolithic but actually chaotic construction of 'poor rural people'. Lay discourses of rural poverty can be further disaggregated so that we can hear the voices of differentially othered poor – for example, the homeless poor, the elderly poor, the young poor, the disabled poor, and so on. Each of these will offer different experiences of poverty, mediating material privation by cultural construction and competence. Indeed, each of these experiences will be cross-cut by other others – gender, sexuality, class, and so on. It is again an important yet difficult task for researchers to ncounter so many worlds of difference. As Blur suggests in Parklife (1994),

> All the people
> So many people
> And they all go hand in hand ...

REFERENCES

Barancik, S. (1990) *The Rural Disadvantage: Growing Disparities between Rural and Urban States*, Center on Budget and Policy Priorities, Washington, DC.

Barnes, T. (1995) Political economy I: 'The culture, stupid', *Progress in Human Geography* 19, 423–31.

Batteau, A. (ed.) (1983) *Appalachia and America*, University Press of Kentucky, Lexington, KY.

Bell, M. (1994) *Childerley: Nature and Morality in a Country Village*, University of Chicago Press, Chicago.

Billings, D. (1974) 'Culture and poverty in Appalachia: a theoretical discussion and empirical analysis,' *Social Forces* 53, 315–23.

Bradley, T., Lowe, P. and Wright, S. (1996) 'Introduction: rural deprivation and welfare', in Lowe, P., Bradley, T. and Wright, S. (eds) *Deprivation and Welfare in Rural Areas*, GeoBooks, Norwich.

Caudill, H. (1963) *Night Comes to the Cumberlands: A Biography of a Depressed Area*, Little, Brown & Co., Boston, MA.

Cloke, P. (1992) 'Rural poverty: some initial thoughts on culture and underclass', in Bowler, I., Bryant, C. and Nellis, D. (eds) *Restructuring Rural Areas*, CAB International, Wallingford.

—— (1993) 'On "problems and solutions": the reproduction of problems for rural communities in Britain during the 1980s', *Journal of Rural Studies* 9, 113–21.

—— (1994) '(En)culturing political economy: a life in the day of a rural geographer', in Cloke, P., Doel, M., Matless, D., Phillips, M. and Thrift, N. *Writing the Rural: Five Cultural Geographies*, Paul Chapman, London.

—— (1996) 'Rural lifestyles: material opportunity, cultural experience, and how theory can undermine policy', *Economic Geography* 72, 433–49.

Cloke, P. and Milbourne, P. (1992) 'Deprivation and lifestyles in rural Wales II', *Journal of Rural Studies* 8, 360–74.

Cloke, P., Milbourne, P. and Thomas, C. (1994) *Lifestyles in Rural England*, Rural Development Commission, London.

Cloke, P., Goodwin, M., Milbourne, P. and Thomas, C. (1995a) 'Deprivation, poverty and marginalisation in rural lifestyles in England and Wales', *Journal of Rural Studies* 11, 351–66.

Cloke, P., Goodwin, M. and Milbourne, P. (1995b) 'There's so many strangers in the village now: marginalisation and change in 1990s Welsh rural lifestyles', *Contemporary Wales* 8, 47–74.

Cloke, P., Milbourne, P. and Thomas, C. (1995c) 'Poverty in the countryside: out of sight and out of mind', in Philo, C. (ed.) *Off the Map: A Social Geography of Poverty*, Child Poverty Action Group, London.

—— (1997) 'Living lives in different ways? Deprivation, marginalisation and changing lifestyles in rural England', *Transactions of the Institute of British Geographers* (forthcoming).

Driver, F. (1988) 'Moral geographies: social science and the urban environment in mid-nineteenth century England', *Transactions of the Institute of British Geographers*, NS 13, 275–87.

Fabes, R., Worsley, L. and Howard, M. (1983) *The Myth of Rural Idyll*, Child Poverty Action Group, Leicester.

Gorham, L. and Harrison, B. (1990) *Working below the Poverty Line*, Department of Urban Studies and Planning, Massachussetts Institute of Technology, Cambridge, MA.

Harrington, M. (1962) *The Other America: Poverty in the United States*, Macmillan, New York.

Katz, M. (1989) *The Undeserving Poor*, Pantheon Books, New York.

Lapping, M. (1992) 'American rural planning, development policy and the centrality of the federal state: an interpretative history', *Rural History* 3, 219–42.

Lewis, H., Johnson, L. and Askins, D. (1978) *Colonization in Modern America: The Appalachian Case*, Appalachian Consortium Press, Boone, KY.

Leyshon, A. (1995) 'Missing words: whatever happened to the geography of poverty?', *Environment and Planning A* 27, 1021–28.

Lohmann, R. (1990) 'Four perspectives on Appalachian culture and poverty', *Journal of the Appalachian Studies Association* 2, 76–91.

McCormick, J. (1988) 'America's third world', *Newsweek*, 8 August 1988, 20–4.

McCormick, J. and Philo, C. (1995) 'Introduction', in Philo, C. (ed.) *Off the Map: A Social Geography of Poverty*, Child Poverty Action Group, London.

McLaughlin, B. (1985) *Deprivation in Rural Areas*, Research report to the Department of the Environment.

—— (1986) 'The rhetoric and reality of rural deprivation', *Journal of Rural Studies* 2, 291–307.

Murdoch, J. and Pratt, A. (1993) Rural studies: modernism, postmodernism and the 'post-rural', *Journal of Rural Studies* 9, 411–27.

—— (1994) 'Rural studies of power, and the power of rural studies: a reply to Philo', *Journal of Rural Studies* 10, 83–7.

Oppenheim, C. (1990) *Poverty: The Facts*, Child Poverty Action Group, London.

Philo, C. (1991) 'De-limiting human geography: new social and cultural perspectives', in Philo, C. (compiler) *New Words, New Worlds: Reconceptualising Social and Cultural Geography*, Department of Geography, University of Wales, Lampeter.

—— (1992) 'Neglected rural geographies: a review', *Journal of Rural Studies* 8, 193–207.

—— (1993) 'Postmodern rural geography? A reply to Murdoch and Pratt', *Journal of Rural Studies* 9, 429–36.

—— (ed.) (1995) *Off the Map: A Social Geography of Poverty*, Child Poverty Action Group, London.

Porter, K. (1989) *Poverty in Rural America: A National Overview*, Center on Budget and Policy Priorities, Washington, DC.

Precourt, W. (1983) 'The image of Appalachian poverty', in Batteau, A. (ed.) *Appalachia and America*, University Press of Kentucky, Lexington, KY.

Rich, S. (1989) 'Ignorance, fear, keep millions off welfare', *Washington Post*, 16 June 1989, p. A1.

Ruggles, P. (1990) *Drawing the Line*, Urban Institute Press, Washington, DC.

Rural Sociological Society Task Force on Persistent Rural Poverty (1993) *Persistent Poverty in Rural America*, Westview Press, Boulder, CO.

Sayer, A. (1994) 'Cultural studies and "the economy, stupid"', *Society and Space* 12, 635–7.

Shapiro, H. (1978) *Appalachia On Our Mind*, University of North Carolina Press, Chapel Hill, NC.

Shucksmith, M., Chapman, P., Clark, G., Black, S. and Conway, E. (1995) *Rural Scotland Today: The Best of Both Worlds?*, HMSO, Edinburgh.

Stephenson, J. (1968) *Shiloh: A Mountain Community*, University Press of Kentucky, Lexington, KY.

Tickamyer, A. and Duncan, C. (1990) 'Poverty and opportunity structure in rural America', *Annual Review of Sociology* 16, 67–86.

Tickamyer, A. and Tickamyer, C. (1987) *Poverty in Appalachia*, The Appalachian Center, University of Kentucky, Lexington, KY.

Townsend, P. (1975) *Sociology and Social Policy*, Penguin, Harmondsworth.

—— (1979) *Poverty in the United Kingdom*, Allen Lane, London.

Walls, D. and Stephenson, J. (eds) (1972) *Appalachia in the Sixties*, University Press of Kentucky, Lexington, KY.

Wildsmith, D. (1990) 'New poor', *Appalachian Journal*, Fall, 94–5.

Wilson, A. (1992) *The Culture of Nature*, Blackwell, Oxford.

Woodward, R. (1996) ' "Deprivation" and "the rural": an investigation into contradictory discourses', *Journal of Rural Studies* 12, 55–68.

Zeller, F. and Miller, R. (1969) *Manpower Development in West Virginia*, West Virginia University, Morgantown, WV.

15

CONCLUSION
Marginality and rural others
Paul Cloke and Jo Little

OTHERNESS AND MARGINALITY

Our main objectives in assembling this collection of essays were essentially twofold. First, we wanted to unpack the increasingly referenced yet still mysterious notion of otherness in the context of rural societies and spaces. Secondly, we were, and remain, deeply curious about the relationship between otherness and marginalisation – that is, how (or indeed, whether) the construction and reproduction of the identities of certain groups and individuals as 'other' is reflected in their marginalisation from the idea and reality of some kind of rural mainstream. While recognising the lack of empirical observation on which to draw and the appeal of simply presenting some detail on the lives of identifiable others, we were con-cerned that this book should go beyond an examination of specific social groups and consider the process by which those groups acquired their marginal status. This involved asking all sorts of questions about the nature of marginalisation as experienced by different people in different places and, in doing this, to engage with a series of theoretical and conceptual discussions on the construction and representation of the rural other.

In Chapter 1 we outlined what we saw as some of the complexities surrounding the notion of the rural other. This discussion has been taken up in a number of the subsequent chapters as different contributors have focused further on the conceptual debate underpinning the construction of the rural other. Thus Jon Murdoch and Andy Pratt (Chapter 3), for example, have discussed the relational nature of 'others' in the positioning of the self and non-self. They also emphasise the importance of ideas of third space and of the need to be open to new categories of otherness which may fall between conventional axes of difference. Other authors, especially Chris Philo (Chapter 2) and Keith Halfacree (Chapter 4), have considered the construction of the other in the context of the evolution and renegotiation of social and economic relations within the countryside, looking at the ways in which 'the other' in rural communities is partly a function of the changing values and expectations of and about rural society.

It is not the purpose of this concluding chapter to revisit earlier discussions on the notion of otherness and the roots of its construction. The majority of chapters in the book, while making reference at some level to the conceptual basis of otherness, have chosen to focus on what could be argued to be the more tangible or practical outcomes of the construction of otherness, namely the state or experience of marginality as captured in the lives of particular groups or individuals. Our hope is that in so doing they have demonstrated that marginality is not simply about the possession or lack of certain essential characteristics (such as whiteness, youth or income) but rather that marginality is dependent upon deeper processes relating to the construction of identities and the positionality of the self and the other. Thus Julian Agyeman and Rachel Spooner's chapter (Chapter 11) exploring the marginality of people of colour within the rural environment emphasises that while racial attack, and feelings of not belonging and/or of hostility, constitute everyday experiences of marginality, the root of that marginality derives from the construction of black identities as other. As these authors then explain, debates on the relationship between notions of 'native' versus 'foreign' as well as on the history of colonialism may be employed to inform ideas about the essential otherness of black identities.

Other examples of this relationship are evident in Annie Hughes' and Jo Little's chapters on rural women. These chapters tell in different ways of the centrality of the domestic lives of rural women and of the marginality of their involvement in paid work. Here it is not women as a group who are marginal – indeed, both chapters show women to be at the core of the village community; rather, because of the otherness of aspects of their female identities, certain parts of their lives (particularly those parts which occupy the public world of waged work outside the village) are experienced as marginal. Understanding that marginality and the power relations behind it requires that we look closely at how gender relations have constructed and represented women's identities as other – and, indeed, how such constructions are spatially and temporally reproduced.

An understanding of the processes of othering is also essential in the appreciation of the fluid and shifting nature of people's marginality. The negotiation and renegotiation of the other in relation to the self means that the experience of marginality cannot be seen as fixed either within or between groups of people or across time and space. So, in crude terms, different people may be excluded from mainstream society at different times in different places and in different social and economic circumstances. This book, therefore, sounds a warning against a rigid and static interpretation of marginality. It is also, however, partly about trying to make sense of some of these differences, to draw lines of similarity which allow us to argue that the particular spatial or temporal patterns do emerge in our attempts to identify marginality.

The explorations in this book of otherness and marginality in a specifically rural context stem from and contribute to our belief that while both
may be seen to be spatially variable, both are to some extent spatially
constituted. Many of the chapters discuss rurality itself, the particular
construction and configuration of social relations in rural areas and their
representation over time, as a feature in the evolution of certain 'other'
rural identities. David Sibley in his chapter, for example, refers to the
interconnectedness of national and rural identities and hence to the view
of the New Age Traveller as other in the context of both 'England' and
'countryside'. David Bell also looks at the meanings and representations
of the rural in the construction of the other. His chapter portrays a very
different yet equally powerful rural identity in the form of the deformed,
imbecilic hick of the rural of US horror films.

The experience of marginality can also be seen as unique to different
people in different spaces and even to different individuals within what are
often taken-for-granted marginal groups such as the poor or the elderly.
Again, however, we would argue that there are clear similarities in their
positions that are spatially constituted. Sarah Harper has shown this to be
the case in her examination of the marginality of the rural elderly, drawing
attention in her chapter to the particular experiences of elderly people in
the countryside and arguing that certain attitudes and characteristics associated with rural communities have combined to shape the nature of their
marginality. Clare Fisher has taken a different tack in examining the
relationship between marginality and space. Her chapter (Chapter 13) on
craft workers has demonstrated the importance of space at a much smaller
scale in a consideration of the spaces of production in the home. She shows
the interconnectedness of space and identity in the context of the daily
lives of craft workers and their occupation of and relationship with different parts of the home and workshop.

Finally, it is important to recognise that an individual's marginality will
also vary in relation to the perception and understanding of the observer
and observed. This seemingly simple point is nevertheless highly relevant
to the experience of and reaction to marginality, as shown by several
chapters. Paul Cloke, for example, in his chapter shows the recognition of
their own poverty by rural dwellers to be remarkably varied. This variation
he sees as crucial to the response to poverty; to the different ways in
which individuals adapt their lives to cope with being poor and to the
attempts by society to deal with that poverty. Owain Jones in Chapter 9
also shows how variation in perception further clouds the relationship
between otherness and marginality. Thus the otherness of rural childhood
is only partially translated into the marginalisation of rural children. Furthermore, that marginality is experienced very differently by children as
children and by adults in response to memories of childhood.

Clearly, the relationship between the process of othering and the experience of marginality is complex and shifting, and rural researchers will need to engage in further theoretical and empirical work to encourage a better understanding of this relationship. We would suggest that particular attention needs to be paid to three important concepts within this work: power, identity and community. All the chapters have referred to at least one of these concepts at some level, but it is perhaps useful here to draw together and reinforce some of the key points that have been raised concerning their centrality within our understanding of the rural other.

Power and identity

While the chapters in this book have focused on the marginality of people they have also stressed the relational nature of that marginality. The web of social relations and power relations which surround individuals and groups in their perceptions and experience of otherness has been shown as complex and shifting rather than one-dimensional and static. The examination of the relational nature of power and marginality has demonstrated the need to think not necessarily of marginalised people but of marginalised identities, and in this way to confront the relationship between subjectivity, hybridity and power.

A focus on identity has allowed greater attention to be paid to the whole question of difference. The recognition that people do not slot into neat unproblematic categories, but rather take on a whole host of cross-cutting attributes, is compatible with the notion of multiple identities. Similarly, the changing nature of power relations is given added weight by an acceptance that identities are not fixed and static, but dynamic and shifting. Below, we consider the formalisation of the relationship between power and identity in the discussion of identity politics, but here we seek to stress the role of debates on power and identity more generally and to emphasise their contribution to our understanding of otherness in rural society.

Work on subjectivity and the self has drawn attention recently to the importance of the body as a grounding for personal identity (Harre 1993), and a wealth of studies has emerged from authors who see the body, as Fiske (1993: 57) argues, as the 'core of our social experience'. Some of the chapters of this book have made reference to such work, demonstrating the centrality of ideas on the body to an understanding of broad notions of identity and otherness in rural society and of more specific and detailed experiences of marginality in rural communities. Sarah Harper's chapter demonstrates the extent to which society's attitudes to ageing and infirm bodies serve to 'other' the rural elderly.

The material circumstances of the body have a significance for the self and for its interaction with the social. As Griffiths (1995) writes, people are born into and grow up in particular bodies. We use aspects of these bodies as points of reference and we have codes to tell us how to respond to and deal with the characteristics they reveal. One direction taken by research on the body has been the examination of exploitation as it relates to the body. Such work has mainly looked at women's bodies, linking exploitation to patriarchy and male control. It has also focused on the body as a site of resistance to conventional power relations, often through the challenge to sexual practice and ideas around sexual morality (see Bell and Valentine 1995; Butler 1990; Lewis and Pile 1996).

While the British countryside may offer relatively few *overt* examples of the exploitation of women's (or other) bodies – in, for example, the sorts of performance associated with carnival events or festivals or acts of prostitution (see Lewis and Pile 1996, but also Smith 1993) – the conventional expectations surrounding the body in rural areas may be just as oppressive. As we have seen, such conventional expectations serve to marginalise particular identities in rural areas – the rural body should be white and heterosexual. It should, as in society generally, be healthy and not in need of assistance for day-to-day routines. Rural women's bodies should be the bodies of mothers or potential mothers; they should not be asexual but neither should they be overtly or erotically sexual. Bodies that do not conform, that have (homo)sexual urges, that are deformed and disabled, are, as David Bell has graphically demonstrated, the stuff of rural horror. They threaten the moral security of the rural community and the order and control of established rural society. Some fascinating work clearly remains to be done on the 'body in the rural' as a site for the constitution and reinforcement of power relations. Equally, we should have concern for the relationship between nature and the body in rural areas. Some feminist work, for example, has considered the power relations which surround the construction of female bodies as somehow more natural and closer to nature than male bodies – linking the woman/man dualism to that of nature/culture (see Rose 1993; Griffiths 1978; Fitzsimmons 1989). Such work has not taken a specifically rural focus, yet it can be seen as one dimension (largely unexplored) of ideas of nature in representations of the rural and of the 'naturalness' of rural women's mothering role.

Beyond discussions of subjectivity and identity, our considerations of power and rural society must also acknowledge the more general importance of representations of rurality. Many chapters noted or discussed the strong images and expectations that surround 'the rural' and which are reproduced through accepted representations of rurality. Such discussions have locked into what has become a flourishing area of academic interest in rural studies, as was pointed out in Chapter 1. Central to work in this area has been a deconstruction of rurality and, specifically, of the notion of a 'rural idyll' (see, for example, work by Bunce 1994; Cloke and Milbourne 1992; Little and Austin 1996). What needs to be stressed here is the degree to which these 'accepted representations' reflect dominant power relations in rural society. They reproduce the preferred composition and values of 'the powerful' in rural society, and in turn reinforce the existing distribution of power and status. At a simple level it is apparent that the dominant representation of rurality is one in which 'the other' is not present (indeed, part of what sustains certain groups of people as 'other' is their continuing absence from the dominant representation). A more detailed reading of the relationship between representation and marginality can show the complex construction of an image of rurality which does not simply mask the existence of those who are 'different' but seeks to marginalise and exclude the lifestyles of those people by presenting them as negative, inappropriate and, in some cases, sinister. It must be recognised that an ability to influence and direct these dominant images (by, if nothing else, simply conforming to the expectations they impose) demonstrates power among the accepted, 'connected' rural inhabitant (or visitor).

IDENTITY, COMMUNITY AND BELONGING

Discussions of marginality – whether of groups or individuals – also have at their core notions of belonging and community. What has been shown in the chapters of this book is that the rural 'other' is effectively denied entry to the established community, at least in terms of the acceptance of particular social identities. We have seen how this denial embraces a set of power relations, and here we argue that it also incorporates moral and ethical issues which link to political questions over representation and citizenship.

One of the most relevant and important of the contradictions which have been shown to exist within contemporary rural society is that between the so-called 'ideal' of the rural community (the warm, tight-knit and accepting community being a central focus of contemporary constructions of rurality), and the recognition of difference. Iris Marion Young has expressed strong reservations about the community as a structure for effective group representation. She is particularly critical of the authoritarianism implicit in notions of 'community', which, she believes, all too often 'serve to iron out multiple identities and impose another kind of oppressive norm' (Phillips 1993: 96). So while 'community' claims to be about tolerance and acceptance – giving the impression of *embracing* diversity – in reality it seeks to deny difference and reject challenge to established norms of behaviour and belief.

This essential contradiction is represented in broader terms by Bauman as stemming from modernity. Society, as he sees it,

> constantly but vainly tries to embrace the unembraceable, to replace diversity with uniformity and ambivalence with coherent and transparent order – and while trying to so this turns out unstoppably

more divisions, diversity and ambivalence than it has managed to get rid of.

(Bauman 1993: 5)

The inability to tolerate difference within and by the contemporary community is reflected in the representation of marginal groups, and challenges, so Anne Phillips contends, the very basis of democracy. Using examples from the feminist literature she draws attention, within the debate on group representation and marginality, to the added difficulty posed by differences within established or definable groups. Here we return to the issues of multiple, shifting and even contradictory identities and of the impossibility of seeing marginality or otherness as something fixed or even embodied in particular individuals. The feminist literature provides a very clear example of the dilemmas involved in attempting to reconcile the recognition of difference with the need for a political identity and group representation. Equally, such literatures emphasise the 'crisis' which feminist writers have encountered in responding to postmodernist concerns with difference and what such directions meant for the feminist 'project'. Such discussions have highlighted the changing role and nature of identity politics in a world where once recognisable 'all-encompassing and rigid social structures like class and the family ... have begun to "detraditionalise" ' (Pile and Thrift 1995: 9).

Many of those writing about contemporary identity politics (e.g. Bondi 1993; Butler 1990; Pile and Keith 1993; Spelman 1990) have questioned the essentialism inherent in the 'classic' political blocs, particularly the feminist movement, of the 1960s, 1970s and early 1980s, demonstrating how new social movements have served to fragment the political universality upon which identity politics rested. The assumption that effective political action can only exist where there is unity has, it is argued, been replaced by the view that political action can and must accommodate a recognition of the shifting nature of identity and of the essential questioning of identity by the individual. Thus, it is claimed that the hybridity of identity need not reduce political power, but may give space to a different kind of power whereby individuals are freed from the inevitability of returning to a particular identity by the acceptance and encouragement of the possibility of constantly redefining and recreating identity. Butler (1990: 15) questions whether the presumption or insistence of unity as a goal is precisely the cause of 'even more bitter fragmentation' among groups. She goes on to argue that unity may set up an

exclusionary norm of solidarity at the level of identity that rules out the possibility of a set of actions which disrupt the very borders of identity concepts, or which seek to accomplish precisely that disruption as an explicit political aim.

278

Also relevant to the debate on identity, as it has emerged through the chapters of the book, is the issue of space. Mention has been made of the shifting nature of identity, and of the belief that identity is not something fixed and final – a personal attribute – but that it may be discovered, formed and re-formed, negotiated and renegotiated. Such a reading of identity suggests that one of the influences on this process of discovery and creation is space. Discussions of identity make frequent reference to spatial metaphors – we talk of the boundaries of identity, of diaspora and of mapping identity. Some authors see this association with space as positive in the sense that it encourages the view that identity is movable, varied and contested. It also raises the possibility of imagined communities and symbolic distance. Others, however, warn that relating identity to space can imply a deadness which conveys only stable and grounded essences (see Smith 1990).

The relationship between space and identity is clearly important in the context of the rural environment. While the majority of chapters in this book have, to some degree, shown how particular identities can be seen as 'other' in contemporary (and in some cases past) rural society, there has also been an attempt by some authors to explore the sense in which the rural itself can foster a particular identity. Clearly, where this has been discussed, care has been taken to avoid the idea of a single rural identity or even multiple such identities to which all rural dwellers ascribe – an idea which would simply reinforce the otherness of those whose identity did not conform to this 'norm'. As we have suggested above, the cultural construction of the rural, and its representation, has been rooted in a set of assumptions, expectations and values. These are not fixed or static, but they do rest on (and in turn reinforce) largely unchallenged meanings and beliefs about the nature of rural life and rural society – meanings and beliefs which are commonly interpreted in relation to place and space. 'The rural' as a place and as a cultural construction is not and cannot be neutral in the creation and expression of identity. Struggles over 'control' of rural space are also struggles – at different levels – over control of the 'meaning' of rurality and of the values that are seen as being represented by rural society and culture.

FUTURE DIRECTIONS IN RESEARCHING THE RURAL OTHER

Finally, we want to comment on the theoretical currency of some of the broader approaches adopted in the study of rural marginalisation. Mention has been made in the introduction to the book and in several of the chapters (for example those by Pratt and Murdoch and by Halfacree) of the 'cultural turn' which has profoundly influenced the direction of rural studies in the UK and has prompted, in particular, a concern with difference and marginality. This expression has been adopted as shorthand to represent an approach or set of approaches that have essentially sought to examine and emphasise the role of cultural factors (in the broadest sense of the word) in explaining the nature and experience of social change in rural society. As was noted in Chapter 1, concern for such cultural factors has been seen as stemming from the exploration of postmodernist perspectives.

There can be no doubt (certainly in terms of the debates raised in this book) that the examination of everyday cultural patterns and practices has provided a new dimension to our research on rural communities and places, and that this new dimension has in turn yielded rich and highly relevant detail to better inform our understanding of the lives of rural people. It has emphasised, in particular, the value of in-depth, local-scale research which attempts to uncover the beliefs, values and perceptions of rural people and, critically, how these contribute to how we know or make sense of the rural. The 'cultural perspective' (for want of a better term) has reaffirmed a belief in the importance of focusing in on everyday practices as a part of understanding rural lives and environments – a belief that was central to the so-called 'community studies' in rural areas in the past yet which got rather lost in the more recent concerns with global structures and processes. At the same time it has, as chapters here bear witness, shown the value of drawing information from the 'ordinary people' of rural areas, recognising the role that such information and such voices contribute to what we know about rural society.

This perspective has, however, also exhibited a number of hazards in terms of the motives for research and the ways in which research is undertaken. Such hazards have not been absent from previous work, but particular dimensions of the study of rural marginality have served to reinforce their importance. Chapter 1 mentioned the problems associated with 'research tourism', using this term to describe a range of dangers connected with the type of research with which we are concerned in this book. Here we want to return to and expand this discussion and apply it specifically to the subject of the rural other.

In Chapter 1 we contextualised contemporary interest in marginalised groups in terms of the development of rural social scientific research. We traced the emergence of a set of theoretical/conceptual arguments in which an appreciation of the existence of 'the rural other' was linked to recent attempts to introduce postmodernist perspectives into rural studies and to inject a cultural dimension in the analysis of the rural economy and society. The point was made that theoretical debate had suffered from, and indeed had contributed to, a serious neglect by rural social scientists of first-hand research on these groups identified as 'other'. Researchers, it seemed, had lacked the empirical currency with which to support theoretical discussion and progression.

We remain desperately short of information on the lives and experiences of a whole range of neglected groups in (or on the margins of) rural society. This book was prompted partly by a perceived need to fill this gap – as has been stressed many times by different authors. It looks, at least in part, to provide a small window on those who live their lives (or part of their lives) outside mainstream rural society and, importantly, outside our representation of that society. While there can be no disputing the need for information to enrich our understanding of the rural other, a number of issues must be borne in mind in the construction and interpretation of that information.

A general concern must be sounded against 'touristic studies' of the rural other. Previous neglect coupled with recent, theoretically informed attention has rendered research on marginalised groups potentially trendy. By the same token, the need for information on such groups has, to some extent, invited opportunistic 'fact-finding' studies. It may be argued, for example, that we have reached a point in rural social studies generally where no study of the rural community is complete without at least passing mention of a list of identifiable others. Care must be taken to ensure that the notion of marginalisation does not simply become a convenient 'excuse' by which to justify narrow description as researchers flit from one 'neglected group' to another.

Equally, it is essential, in order to avoid glib and even potentially damaging research of such a touristic nature on the rural other, that (i) we observe certain codes regarding the sensitive handling of research on marginalised groups, and (ii) our work on such groups and individuals is at some level linked into work on the process of marginalisation and the power relations embedded in the construction of otherness.

This first issue of sensitivity demands careful consideration of the relationship between researcher and subject. A key concern here is the identification of marginal groups; frequently such 'groups' are not defined as such nor perceived as being marginal or neglected, either by themselves or by the rest of rural society. In addition, anonymity remains the most powerful weapon of some (particularly gay and lesbian groups) whose otherness can engender extreme prejudice and violence as well as simply a lack of appreciation and understanding of their lifestyles. Researchers may find frequent contradictions between the desire to raise awareness of the experiences and treatment of marginalised groups and the need to preserve confidentiality and protect anonymity. As demonstrated by the different approaches adopted by the studies included in this book, there are clearly few general 'rules' that can be applied, but quite simply, the imperative of not allowing academic goals to override personal individual rights and security must be upheld.

A closely related issue here concerns the broader purpose of research on the rural 'other'. One of the more disturbing aspects of research tourism, and one that is particularly pertinent to the study of marginal groups, is the absence of any tangible output from the research process which is of relevance, use or interest to the subject(s) of that research. In other words, a timely piece of research may be used solely for academic gain with little or no thought about either the 'ownership' of information or the potential of research to provide positive gains for marginalised groups (who, by definition, generally have limited access to information and power). This raises a whole set of ethical issues around research on the marginal and powerless in society which are taken up below. It is of crucial importance, however, that such issues are seen as relevant to the questioning of depth and sensitivity of research practice and the criticism of research tourism.

The chapters contained in this book show how the study of neglected 'others' can broaden our understanding of rural society, reinforcing the argument that while we should take care not to indulge in research tourism, we should not be deterred from seeking to learn more about the lives of those living beyond mainstream rural society. The chapters also demonstrate, however, that researching the rural other, at whatever level, involves us in an examination of the distribution and operation of power relations among rural people and rural institutions. Understanding such power relations requires us to look not merely at the day-to-day experience of marginality, but, through those experiences, at the processes through which people become marginalised.

Notwithstanding the dangers in the study of rural marginality, this new dimension which has been prompted and encouraged by the cultural 'turn' has clearly been widely recognised as providing a positive direction to rural social scientific research. It has not only promoted discussion on the specificity of people's lives, feelings, emotions and beliefs, previously ignored or undermined, but also (again, as has been seen throughout the book) opened up broader theoretical debate, injecting into rural studies ideas from a range of disciplinary backgrounds. It is also clear that considerable scope remains in terms of the development of cultural perspectives in rural research. The chapters here may have introduced new and exciting information on the lives and experiences of rural people, but they have also demonstrated something of the paucity of real understanding that exists. With all the caution and caveats expressed in the earlier part of this chapter, it *is* important that the sort of information suggested by a cultural approach continues to be constructed in rural communities.

While we do not wish to deny or detract from the obvious value and contribution of cultural or postmodern perspectives on rural research, it is important to register some concerns about the wholesale adoption of such approaches. Our discussion on research tourism has articulated some of the dangers (as we perceive them) which can be seen to surround the implementation and purpose of research on the rural other – but we also want to broaden such concerns to incorporate the theoretical dimension of research on these issues. Our main concern in this respect relates to the downplaying or even ignoring of other perspectives – particularly political economic perspectives – in the adherence to a cultural approach.

The debate over the role and value of different theoretical perspectives in the explanation of social and economic change in rural areas is not one to be resolved here. Our point is rather to argue, at a general level, that research needs to remain open to different perspectives. At a more specific level in relation to the subjects examined in this book, our concern is not to discard some of the central strengths of political economic approaches as they have been (and continue to be) developed in rural research. It is important, for example, that we do not dismiss out of hand the work that has been done from a political-economic direction on global processes of restructuring and that has, crucially, situated rural economic relations within a wider examination of the capitalist transformation of production. Such work seems to us to have been extremely valuable in terms of the identification and exploration of the broad shift from production to consumption in rural areas, which in many ways provides an essential backdrop to cultural analyses. It is also critical, in a similar vein, that we do not lose sight of some of the broader axes of social structure and social relations, particularly class and gender, which still constitute areas of considerable inequality in contemporary society. This is not to say that these social divisions should be treated as unproblematic, but rather that we should not underplay their continuing relevance to patterns of deprivation and marginalisation in the rush to focus on human agency and develop cultural forms of explanation.

A further concern in the neglect of political economic approaches is the potential loss of an ability to question and challenge social inequality in rural communities. This is clearly not an inevitable result of either the adoption of cultural perspectives or the rejection of political economy. It is the case, however, that with the identification of patterns and process of social deprivation in the context of debates over class relations and change, the political economic perspective has often encouraged the search for equality and the alleviation of deprivation. Equally, an emphasis on culture and its essential focus on difference, and the less tangible concerns with, for example, representation and identity, can in some cases detract from the purpose of challenging social inequality.

The appeal for continued recognition of political economy approaches should not be seen as a wish to somehow undertake a merger of the political economic with the cultural. It may be that the two approaches need to run side by side – that their very different contributions can be acknowledged but not somehow diluted in an attempt to make them mutually acceptable. Elements of both approaches are oppositional, but only through recognising and debating such unconformities can we advance the real worth of each approach.

The chapters of this book have used a variety of perspectives to relate some colourful stories of the lives of some of those living in rural areas. An attention to what we might term 'cultural geographies' has provided a richness to these stories which has largely gone unrecognised in the past. While this work has helped to inform us in areas characterised previously by neglect, a recognition of the ethical questions surrounding studies of 'the other' stresses that we should be more than simply informed. We need to interact with the stories we tell and with the lives of those who we write about. We should not, like tourists, return to our own place in the world with a sense of the exotic, but without our own lives being significantly changed at all (Griffiths 1995).

Our hope in publishing this book is not only to engage with the lives of rural people but also to influence the future directions of rural research. In gathering information on rural marginalisation we have come to realise just how much is left unsaid and unwritten. There remains considerable scope for further work on the experiences of those outside mainstream society. What is also clear, however, is that considerable opportunity remains for the development of theoretical and conceptual research on constructions of, for example, difference, identity and representation and how these relate to notions of power in the rural environment.

REFERENCES

Bauman, Z. (1993) *Postmodern Ethics*, Blackwell, Oxford.

Bell, D. and Valentine, G. (1995) 'Queer Country: rural lesbian and gay lives', *Journal of Rural Studies* 11 (2): 113–22.

Bondi, L. (1993) 'Locating identity politics', in Keith, M. and Pile, S. (eds) *Place and the Politics of Identity*, Routledge, London.

Bunce, M. (1994) *The Countryside Idyll: Anglo-American Images of Landscape*, Routledge, London.

Butler, J. (1990) *Gender Trouble: Feminism and the Subversion of Identity*, Routledge, New York.

Cloke, P. and Milbourne, P. (1992) 'Deprivation and lifestyle in rural Wales', *Journal of Rural Studies*, 8 (4): 359–71.

Fiske, J. *Power Plays Power Works*, Verso, London.

Fitzsimmons, M. (1989) 'The matter of nature', *Antipode* 21: 106–20.

Griffin, S. (1978) *Women and Nature: The Roaring Inside Her*, Harper Colophon, New York.

Griffiths, M. (1995) *Feminisms and the Self: The Web of Identity*, Routledge, London.

Harre, R. (1993) *Social Being* (second edition), Blackwell, Oxford.

Keith, M. and Pile, S. (eds) (1993) *Place and the Politics of Identity*, Routledge, London.

Lewis, C. and Pile, S. (1996) 'Women, body, space: Rio carnival and the politics of performance', *Gender, Place and Culture* 3 (1): 23–42.

Little, J. and Austin, P. (1996) 'Women and the Rural Idyll', *Journal of Rural Studies* 12 (2): 101–11.

Phillips, A. (1993) *Democracy and Difference*, Polity, London.

Pile, S. and Thrift, N. (1995) *Mapping the Subject: Geographies of Cultural Transformation*, Routledge, London.

Rose, G. (1990) *Feminism and Geography*, Polity, London.

Spelman, E. (1990) *Inessential Woman: Problems of Exclusion in Feminist Thought*, The Women's Press, London.

Smith, N. (1990) *Uneven Development* (second edition), Blackwell, London.

Smith, S. (1993) 'Bounding the borders: claiming space and making place in rural Scotland', *Transactions of the Institute of British Geographers* 18 (3): 291–308.

INDEX